Praise for
The Lean Farm Guide to Growing Vegetables

"Ben Hartman and I share similar approaches to growing vegetables; we both run highly productive farms using efficient techniques and well-designed space and procedures. With this book, however, I believe Ben has taken our craft to new levels with fresh ideas and different strategies. The information here provides incredible value for any small-scale farmer seeking a compact, yet profitable farm model. I highly recommend this book."

—JEAN-MARTIN FORTIER, author of *The Market Gardener*

"All revolutions require a leap in consciousness plus a set of daily practices to sustain and gain the full benefits of the new consciousness. Ben Hartman is that rare person who could describe the lean farming revolution (in *The Lean Farm*) and then provide proven practices from his own farm in his new book, *The Lean Farm Guide to Growing Vegetables*, to help fellow revolutionaries grow good food in a lean way for the long term. Together these volumes describe all you need to know for a sustainable lean revolution on your farm. The rest is up to you."

—JIM WOMACK, founder and senior advisor,
Lean Enterprise Institute

"This manual of growing follows the lean principles of Ben and Rachel's farm, its pages are jam-packed with useful advice for efficient organic growing. 'Lean' is a great paradigm for directing your time and energy into a fully effective approach, whether in a large garden or small farm. I was fascinated to read all the examples of seeding, planting, and harvesting, and the gorgeous photos show the high productiveness of these methods."

—CHARLES DOWDING, coauthor of *No Dig Organic Home & Garden*

"Everyone strives for efficiency in vegetable farming, but Ben Hartman has actually achieved it. In his lean farming books, he provides a clear-headed approach to achieving efficiency of space, time, and resources. Every vegetable farmer who wants to be profitable *and* enjoy the farming life would do well to read these books closely."

—LYNN BYCZYNSKI, author of *Market Farming Success*
and *The Flower Farmer*

"Ben Hartman is a true innovator for the small farm. *The Lean Farm Guide to Growing Vegetables* takes the lean techniques from his first book even further with new innovations and greater detail. I have been eagerly awaiting this addendum so that I can start trying these practices on my farm. This book is a must for any market grower who wants to push the boundaries of profitability while finding life balance at the same time." —CURTIS STONE, author of *The Urban Farmer*

"Ben Hartman clearly illustrates how the continual process of lean thinking can benefit every farm. From targeted market planning driven by what customers truly value, to effective and efficient production from planning through harvest, and ultimately to putting money into your bank account, this book is full of practical and inventive ideas that will help your farm prosper."

—RICHARD WISWALL, author of *The Organic Farmer's Business Handbook*

"This is a great book on the methods that make Clay Bottom Farm a successful compact farm, explaining the thinking behind those methods. Ben gives clear, detailed descriptions of the day-to-day systems, and further illustrates the concepts he first laid out in *The Lean Farm*. This companion, *The Lean Farm Guide to Growing Vegetables*, reveals a refined system that makes a great jumping off point for any aspiring grower, and gives seasoned growers ideas for how to improve their own systems."

—JOSH VOLK, author of *Compact Farms*

"Another gem from Ben Hartman. *The Lean Farm* introduced a dynamic new way of thinking about small farm businesses, pointing us in the right direction. Now, *The Lean Farm Guide to Growing Vegetables* gives us the detail, specifics, and tricks of the trade that show us how to execute lean strategies on a small farm. Ben is a key entrepreneur in diverse, sustainable, small farm agriculture, and is emerging as the Midwest's heir apparent to Eliot Coleman. *The Lean Farm Guide to Growing Vegetables* is an indispensable resource for all small farmers, new or experienced, young or old."

—STEVE HALLETT, professor of horticulture, Purdue University; author of *The Efficiency Trap*; coauthor of *Life without Oil*

"Applications of lean principles to a new arena are often confusing and shallow. *The Lean Farm Guide to Growing Vegetables* is clear and deep. This book is based on real-world experience, drawing on lean principles to develop a radically new approach to farming that gets you more for less."

—JEFFREY LIKER, author of *The Toyota Way*

"I'm an impatient reader but always make time for Ben's writings. Every time I found myself wondering about a nugget of detail in *The Lean Farm Guide to Growing Vegetables* it would appear within the next paragraph or two. This book fully explores production details that most authors skim over, and in vegetable production—as in any craft—details matter."

—PETE JOHNSON, founder, Pete's Greens, Craftsbury, Vermont

"If you liked *The Lean Farm*, you'll love *The Lean Farm Guide to Growing Vegetables*. Ben's first book was the big picture of what lean principles are and how they can be used to make farms more efficient. This book focuses in for the close-up, with specific examples of how to apply lean principles to vegetable farming and case studies from the author's own farm experience. Ben's discussion of *kaizen*, the practice of continuous improvement, reminds us we can all be more efficient. Read this book with last season in mind and you'll be inspired with ideas of how to streamline next season. After reading this I have more than a few lean ideas I'm going to apply on my own farm."

—ANDREW MEFFERD, author of *The Greenhouse and Hoophouse Grower's Handbook*

The
Lean
Farm
Guide to
Growing
Vegetables

The Lean Farm

Guide to

Growing Vegetables

More In-Depth
Lean Techniques for
Efficient Organic Production

Ben Hartman

Chelsea Green Publishing
White River Junction, Vermont

Project Manager: Angela Boyle
Project Editor: Michael Metivier
Copy Editor: Katherine Scott
Proofreader: Laura Jorstad
Indexer: Linda Hallinger
Designer: Melissa Jacobson

Printed in the United States of America.
First printing October, 2017.
10 9 8 7 6 5 4   22 23 24 25 26

Our Commitment to Green Publishing

Chelsea Green sees publishing as a tool for cultural change and ecological stewardship. We strive to align our book manufacturing practices with our editorial mission and to reduce the impact of our business enterprise in the environment. We print our books and catalogs on chlorine-free recycled paper, using vegetable-based inks whenever possible. This book may cost slightly more because it was printed on paper that contains recycled fiber, and we hope you'll agree that it's worth it. *The Lean Farm Guide to Growing Vegetables* was printed on paper supplied by Versa Press that is made of recycled materials and other controlled sources.

Library of Congress Cataloging-in-Publication Data
Names: Hartman, Ben, 1978– author.
Title: The lean farm guide to growing vegetables : more in-depth lean techniques for efficient
  organic production / Ben Hartman.
Description: White River Junction, Vermont : Chelsea Green Publishing, [2017]
  | Includes bibliographical references and index.
Identifiers: LCCN 2017027614| ISBN 9781603586993 (pbk.) | ISBN 9781603587006 (ebook)
Subjects: LCSH: Vegetable gardening. | Organic gardening.
Classification: LCC SB324.3 .H375 2017 | DDC 635—dc23
LC record available at https://lccn.loc.gov/2017027614

Chelsea Green Publishing
85 North Main Street, Suite 120
White River Junction, Vermont USA

Somerset House
London, UK

www.chelseagreen.com

Contents

Preface

The greatest fine art of the future will be the making of a
comfortable living from a small piece of land.

—Abraham Lincoln

On February 19, 1909, Franklin Hiram King, an agronomist with the
University of Wisconsin, boarded the steamship *Tosa Maru* in
Seattle, Washington, and embarked on a nine-month journey across
rural Japan, Korea, and China. It was the first trip of its kind, a journey to
the remotest parts of the East to learn how farmers on small plots of land for
generations have managed to feed millions of people without depleting
their soils.

King waded through rice paddies, strolled around vegetable plots, and
talked to countless peasant growers. Their secret, it turns out, was simple:
waste nothing. "Manure of all kinds, human and animal," King wrote, "is religiously saved and applied."[1] Along garden edges King saw rows of receptacles
for collecting liquid manure and piles of ashes and manures for fertilizing. He
saw compost pits dug in front of houses, almost as showcases of family life.
Everything from old wooden bowls, chopsticks, even worn-out clothes and
trodden dirt floors was reused, composted, and applied to the land. One
farmer told him, "We throw nothing away. It is worth too much money."[2]

The waste-free farming—or *living*—that King saw was not unique to the
East, of course. My ancestors in Germany hung laundry out to dry over their
produce patches so plants could use the dripping water. Decades ago Pari-
sian market growers carried baskets on their backs containing horse manure
from the city center to their gardens. George Washington Carver, the great
botanist and inventor born into slavery, spent his career imploring poor
farmers to turn waste into "useful channels."

More now than ever we live in a throwaway culture. It's hard to imagine
a return to such waste-free living. But we can do much better. Somewhere
between the industrial revolution and the advent of get-big-or-get-out agriculture, between the "green" revolution and the rise of gigantic-scale farming,
the *waste nothing* approach got lost, discarded like a stack of old scythes.

It is time to bring it back.

Garden plot of my ancestors in Germany, where laundry was hung out to dry over vegetables.

In my first book, *The Lean Farm*, I discussed how any farm can use the lean system—a production scheme popularized in Japanese industry—to cut out waste, increase efficiency, and become more profitable with less work. This book is a field guide companion to *The Lean Farm* for market vegetable growers. I will show in greater detail how we have implemented lean thinking in every area of our work at Clay Bottom Farm, in Goshen, Indiana. For example, in *The Lean Farm* I mention the idea of *kanbans*, or replacement signals, to maximize land use; here I show you our *Kanban Maps*. In *The Lean Farm* I mention our use of germination chambers to reduce defect waste; here I explain how to build one. The lean system allows us to earn a comfortable living on an acre of land, and this book shows just how we do it. Think of this as a how-to manual, with a lean twist.

Acknowledgments

Many hands helped create this book. Stephan Rauh, an intern on our farm in 2016, typed out notes I dictated for this book while I worked. Stephan also read lengthy portions of the manuscript and offered invaluable feedback from a first-time-farmer perspective. Thank you, Stephan. Throughout that growing season I frequently passed around our phone to workers and asked them to take pictures of our systems. Thank you to Sophie Lapp, Anna Nafziger, Joshua Jantzi, Philip Mann, and everyone else who helped document our farm's processes.

I sent my brother, Fritz Hartman, the director of the Good Library at Goshen College in Goshen, Indiana, on a quest to find guidebooks used by premodern farmers in Japan. After an arduous search, Fritz discovered a few scrolls and manuscripts, one dating to before 1000 CE, and supplied me with valuable academic research on the topic. Thank you, Fritz.

Emma Gerigscott, an artist and former Clay Bottom Farm intern, created the illustrations in this book. Dr. Ervin Beck, professor emeritus of English at Goshen College, edited the manuscript and leaned it up. Dr. Vern Grubinger, berry and vegetable specialist with the University of Vermont Extension, and Clare Sullivan, of the Oregon State University Extension for Small Farms and Community Food Systems, reviewed chapters 2 and 3 and offered advice that improved those chapters. David Johnson contributed professional photos. Ben Watson, Angela Boyle, Michael Metivier, and the entire team at Chelsea Green cheered me on and made this book look great. Thank you, all.

The biggest thank-you goes to my wife, Rachel Hershberger, and to our boys, Arlo and Leander. Rachel combed the manuscript with a deft touch and improved it and gave me time to write. The boys frequently played at my ankles as I wrote.

Hosing root vegetables.

Five Steps to a Lean Vegetable Farm

I think farms . . . are kind of like two-year-olds.
They're very loud and very insistent about what they
need and what they want from you. If you don't set some
limits, you're going to be a slave to the two-year-old.

—CHRIS BLANCHARD

There's a crack in everything—that's how the light gets in.

—LEONARD COHEN

I

On a late Friday afternoon in the winter of 2006, I had finally had enough. I packed up my drywall knives, washed my plaster-covered hands, and walked out the front door of the large house I had been drywalling for the past two months.

I hadn't always worked in construction. I grew up on a 450-acre corn and soybean farm in northeast Indiana. As a teenager I drove tractors for my dad, hauling in wagons from the field or chisel-plowing after the harvest. I always had a passion, though, for vegetables, which were not a part of the farm's income. In our home garden, after school and during the summers, I helped my mom trellis tomatoes, haul in watermelons, and till in leaves in the fall. In high school I started a garden club that grew and sold food to our cafeteria. In college I grew a banana tree under grow lights (the fruit never ripened), raised chickens in an abandoned car, and ran a twelve-member CSA (community-supported agriculture).

My wife, Rachel, grew up in town but with a large garden as well. She spent summers with her family growing beans and tomatoes in a large garden and canning food for winter. After graduating from college we both worked at Sustainable Greens, an organic farm run by Kate and James Lind, where we learned the difference between Rouge d'Hiver and Lolla Rosa lettuce, and how to protect spinach under greenhouses for winter harvests. We lived in a communal house, where everyone shared income from a market garden.

In a few years we married, bought a house in town, and farmed our yard, plus a few rented patches about town, while I took on construction work and Rachel worked in the middle school. That first season we earned $5,000 growing beans, tomatoes, peppers, onions, and greens for local customers. The next season we earned a little more. We began to wonder, what if we quit our jobs and farmed full-time? Could we make it? We had caught the farming bug! That fall I stopped drywalling for good. A few weeks later, with the help of a family loan, we bought a 5-acre farm 6 miles from town, sold our house, and started a new life.

▪ ▪ ▪

Meanwhile, across town, after swinging a hammer all afternoon in 20° weather, a local contractor named Jason Oswald decided he had had enough, too. Jason was known for his physical prowess: he once caught a 4-by-12-foot piece of drywall midair as it fell from a window above him. But at 35 he was tired of frozen toes. His dream was to start a bar. Not a dingy drinking hole, but an upscale pub with local food, craft beers, and a clean atmosphere.

That summer Jason rented a vacant storefront on Main Street and dug in. He poured a concrete countertop, installed restaurant booths, set up beer taps, built a kitchen, and hired a chef, bartenders, and wait staff. His bar and restaurant, Constant Spring, became our first wholesale customer.

▪ ▪ ▪

We decided not to grow much food that first year on our new farm. Instead we focused on infrastructure. We installed drain and water lines, built a greenhouse, and remodeled an old milking parlor into a processing room. We planted a fruit orchard and raspberry plants. We went to farm auctions and bought a 1949 Ford 8N tractor, a two-bottom plow, and hand tools. We plowed our ground and seeded vetch, clover, and rye. The soil was heavy clay. We fenced off 2 acres for steers and chickens.

In the second and third years, we were off and growing. Our farm was a frenzy of high-energy experimentation. We hosted interns from California, Georgia, Pennsylvania, Missouri, Louisiana, and Indonesia. We grew turmeric, Chinese ginger, and heirloom corn. We planted horseradish and figs.

Complexity is the enemy of lean. We are always on the lookout for ways to simplify our work.

We built a kiln and made dinnerware out of our clay soil. We built tools and temporary greenhouses. We stuffed our barns full of half-finished projects.

Our biggest project: we bought two greenhouses at an auction an hour south of our farm. Along with friends, we spent a month moving them piece by piece. We stashed nuts and bolts in yogurt cups and wrapped up metal framing parts with stretch wrap.

We put up the first greenhouse in August. Hasty in our construction, we did not pour enough cement. Three weeks later, a heavy wind pried the greenhouse loose from wet soil, slamming it into our barn. We heard the *thunk* from inside the house. That evening, with help from Amish neighbors who arrived within 10 minutes, I climbed up onto the barn and cut large slits into the plastic so the greenhouse wouldn't sail to Michigan. Then I headed inside and walked straight to the refrigerator. I opened a beer, sat at the kitchen table, and stared at a crack in the wall.

Back at Constant Spring, business was booming. Goshen had been ready for a new type of bar, one with hip décor, craft beer, and local food. Jason's timing had been spot on. He called me. "Ben, what would it take for you to supply us with tomatoes?" He told me the amount they used.

"A thirty-by-ninety-foot greenhouse," I replied.

"You buy it, I'll help put it up," he told me. "And I promise I'll buy the tomatoes you grow in it." This is, in microcosm, the story of the local food movement: local farmers and businesses partnering to create an alternative to shipped-in food.

While our farm had its share of setbacks, there was no time to mope. Within a week an anonymous customer who had heard our greenhouse-on-the-barn story sent us a check to help us cushion that blow. And we set to work with Jason building his tomato greenhouse. By the next year, we were stocking the Constant Spring with tomatoes from May to November. Within two years two other artisan restaurants started up in Goshen, and our CSA doubled in size. On the farm, we were putting in place systems to meet the demand.

But something needed to change. The greenhouse on the roof was a symptom of deeper problems, an outgrowth of our farm's chaos—of a lack of focus, of too much stuff and too many projects. If we didn't change course and adjust how we farmed, we would face more calamities and our reserve of new-farmer grit would run out.

II

In 2011 I received an email from Steve Brenneman, a CSA customer and the owner of an aluminum trailer factory. He had recently transformed his factory using the lean system, a Japanese invention designed to cut out waste and increase profits with less work. Brenneman offered to help us do the same on our farm. After initial skepticism (factory methods on farms haven't always churned out sustainable outcomes), we took him up on his offer. Over the following several seasons, we proceeded to declutter our workspaces and cut waste from our processes. We now earn a comfortable living from our farm, and neither of us works off the property. We have stable, long-term customers, and our work doesn't overburden us. I tell this story in my first book, *The Lean Farm*.

Five core lean principles have benefited us the most. These principles underlie the production systems in this book, and we use them to this day.

1. Keep only the tools that add value.
2. Let the customer define value.
3. Identify the steps that add value.
4. Cut out the *muda*–anything that does not add value.
5. Practice *kaizen*–continuous improvement.

Lean Math

We agreed to give lean a try because we came to understand a simple formula:

Eliminated waste = capacity

Our business was growing, and we needed more capacity, but we were tired of sinking money into the farm. This formula offered an out: we could grow our business—even free up time—without more investment. We just had to get rid of the waste.

Here is how it works. If you find a way to shave just one hour per week off your work, over a period of 20 years you will have freed up six months of time. Shave four hours per week and you free up an entire year every 10 years.

This "free" capacity is yours to do with what you want. You can expand and produce more products with the freed-up time. You can also take time off. In 2015 we took five three-day camping trips in the middle of the growing season. In 2016 I took two mornings off per week to spend with our two sons, ages two and one. While there are times when all farms need to invest, getting bigger all the time isn't the only way to grow a business.

1. KEEP ONLY THE TOOLS THAT ADD VALUE

Lean relies on an organization method called 5S for keeping workspaces clean. The most important step is *seiri* (sorting), or discarding any item from your workspace that is not used every season to create value. This has not always been easy for us to do. At first we placed too much stock in tools that, realistically, we would never use. Now we have routines for constantly making sure we are surrounded only by what we need.

Farmers love to collect tools and hoard things. But we must check that instinct because everything we keep has a cost. It costs time to trip over shovels and look for hoes. It costs to store items—property taxes on buildings and expenses to maintain those buildings. Plus there is the mental cost of living with clutter. These are costs that suck energy and money without giving anything back besides the vacuous satisfaction of ownership.

Lean methods guide decision-making in every aspect of our farm, from seeding to selling.

We sort tools twice each year, discarding those we don't use and placing our favorites in easy-to-reach locations, close to their points of use.

2. LET THE CUSTOMER DEFINE VALUE

A second practice underlying our daily work is to always have our customers and their wishes in mind. We are growing food for them, and the products they want, both goods and services, guide our work. Specifically, we seek out answers to these three questions:

What do our customers want?
When do they want it?
What amount do they want?

The more precisely we answer these questions and deliver on the answers, the more profitable we will be at year's end. To be sure, we don't fulfill every wish. For example, while customers might love watermelons, we don't grow them because, for us, they are not profitable. Still, continually posing and answering the three questions ensures that paying customers, not our own whims, steer our farming.

3. IDENTIFY THE STEPS THAT ADD VALUE

The next principle we follow is to carefully trace value. In which steps do tomatoes and peppers and carrots become more valuable? When are we adding to their value, and when are we just spinning wheels? In a factory setting, value might be added when steel is welded to steel. That is the point when each piece is suddenly worth more. On the farm we've found that surprisingly few actions—seeding, harvesting, and washing food, for example—build value. These direct actions cause our food to be worth more.

4. CUT OUT THE *MUDA*— ANYTHING THAT DOES NOT ADD VALUE

According to the lean system, actions on your farm that don't directly add value—like lawn mowing, cleaning, and leafing through catalogs—are considered *muda*. The closest English equivalent is "waste." The Japanese managers who developed the lean system were precise when outlining forms of *muda*. Here they are as applied to the farm:

1. *Overproduction*—spending resources on products that do not sell
2. *Waiting*—products or people sitting idle
3. *Unnecessary transportation*—too much driving, too little farming
4. *Overprocessing*—washing and packaging more than the customer requires
5. *Too much inventory*—unused supplies or sellable food plugging up the farm
6. *Unnecessary motion*—any physical action that is not creating value
7. *Making defective products*—twisted carrots, wormy tomatoes
8. *Overburdening* (muri)—unreasonable effort to get a job done
9. *Uneven production and sales* (mura)—peaks and valleys rather than a smooth, predictable workflow
10. *Unused talent*—a good idea that went unspoken

Lean Does Not Mean Cheap

Hear the word *lean* associated with vegetables and you might imagine growers cutting corners to get jobs done faster. However, lean is not only about cost cutting. More fundamentally it is about discovering what customers value and pivoting your actions toward that value. As Eliot Coleman writes in *Four Season Harvest*, "Food is not a commodity to be produced as cheaply as possible. It is the living matter that fuels our systems."[1] When used correctly, lean empowers a grower to produce high-value foods with high profit margins, instead of a race to the bottom, producing the cheapest food on the market.

On our farm we take *muda* elimination seriously. We look for the 10 *mudas* all the time. When we see them, as much as possible, we get rid of them. While *muda* is usually translated as "waste," not all *mudas* are completely superfluous. We still file taxes, occasionally pull weeds, and send invoices to customers, even if these actions do not directly cause our products to be worth more. Not all *mudas* should be completely eliminated. The goal, rather, is to minimize the amount of time and energy given to non-value-adding activities.

We've come to realize that every activity we perform is either *muda* or value adding. There are no exceptions. Our job is to see the difference, and steer toward value. If we find ourselves performing a *muda* task, we ask, how can we shorten it or, better yet, eliminate it? As Shigeo Shingo, a Japanese engineer and arguably the world's leading expert on manufacturing efficiency, observed, only the last turn of a bolt tightens it—the rest is just movement. We want to focus on bolt tightening, so to speak—on seeding, washing, harvesting, and selling—and minimize everything else. Cutting out *muda* has been a powerful force on our farm, increasing our profits every year, while we work less.

5. PRACTICE *KAIZEN*— CONTINUOUS IMPROVEMENT

The final step is to practice *kaizen*, or continuous improvement, rooting out more waste every year from the farm, and aligning production more tightly with customer demand. We have adopted the philosophy "When you fix it, fix it again."[2] Our goal is to create a farm culture where everyone on the team joins together in an effort to achieve zero-waste production.

The end goal of the five steps is compression. We want to shorten the timeline from beginning to end, from the idea to grow a crop to the moment we receive cash. Along the way our production area shrank from 5 acres to less than 1, and we shed thousands of pounds in tools and supplies. Our process timeline shrank, too, as we simplified our work in an effort to focus on value-adding tasks. Taiichi Ohno, a Chinese-born engineer considered to be the father of the Toyota Production System (the precursor of the more generalized lean system), summarized lean this way:

> *All we are doing is looking at the time line from the moment the customer gives us an order to the point when we collect*

the cash. And we are reducing that time line by removing the non-value-added wastes.[3]

Here are the steps in the timeline of crops at Clay Bottom Farm:

1. Receive orders and plan the year.
2. Prepare the fields.
3. Make compost.
4. Start seeds in the propagation greenhouse.
5. Transplant or direct-seed.
6. Control for weeds and pests.
7. Harvest, wash, and package.
8. Deliver to restaurants or CSA customers, or display at market, and collect cash.

In this book we will travel together along the journey of this timeline. I describe how we have compressed each step to increase our hourly wage and free up time and energy for our children and other pursuits—to achieve that elusive work-life balance.

We started farming because we loved working with plants. Lean has not stripped the joy from our work. It has given us focus so that with fewer distractions—less *muda*—we enjoy our work now more than ever.

While in the pages that follow I show you the particular techniques we use, my intent is not to convince you that this is the best way to farm. I do not think that there is such a thing as a model farm—nor should there be. If someone comes along trying to convince you that one particular technique is your route to millions, they are wrong. Hold on to your suspicions. There are many ways to grow food successfully. The pages that follow merely reveal a snapshot of what works for us in our context. They show where we have landed in our production at this time—after years of thinking lean.

Photo courtesy of David Johnson Photography.

PART I

Leaning the Timeline

CHAPTER 1

Planning the Year with *Heijunka* and *Kanban*

> *How we spend our days is, of course, how we spend our lives.*
>
> —ANNIE DILLARD

In early January of the second winter on our farm, after we'd tucked away the Christmas tree and cleaned up from a New Year's party, Rachel and I gathered our laptops, stacks of blank paper, drafting rulers, seed catalogs, and a file full of farm planning charts that we'd cut out of magazines or copied from books. We spread it all out on the dining-room table. It was time to plan our year.

We flipped through seed catalogs, keeping a list of varieties we might want to try. Then we created an Excel spreadsheet, with rows listing our crops and columns like "time to seed," "spacing," "customers who want this," "fertility plan," and so on. We drew maps of the garden, plotting precisely where each seed should go. We created profit calculator templates, sales projection spreadsheets, and day-by-day to-do lists. We were ambitious. The whole process took weeks.

In retrospect, we created more than we needed. While we enjoyed this work back then, now we want time to spend on other things. We have since leaned up planning, which now takes just a few hours. First, we create a simple seeding calendar, which we call our *Heijunka* ("load leveling") Calendar. Next, we create field maps that tell us approximately how many beds of each crop to plant. We call these *Kanban* ("replacement") Maps. (I will explain the concepts of *heijunka* and *kanban* below.) Both documents are based as much as possible on advanced orders and data-driven projections, not hunches as to what might sell.

The key point in planning a farm season, we have learned, is to figure out the information we need, and then to assemble it as efficiently as we can. We don't have to chart out every minute of our year, or plan for every square inch; overdoing is a form of waste. We won't receive a check in the mail just because we spent the afternoon leafing through seed catalogs or analyzing records. With quick planning, we can focus our efforts on seeding, harvesting, and selling—the steps that build actual value.

The *Heijunka* Calendar

The *Heijunka* Calendar tells us when to seed each crop, based on our orders. The *Kanban* Map (next section) tells us how much to seed.

In the first few seasons I recommend using stakes to quickly track seeding and harvest dates as you develop a seeding calendar specific to your bioregion. On this stake "S" stands for the seeding date and "T" represents transplant date. The dates underneath indicate harvests. On the backside I jot notes, such as "bolted July 10" or "try seeding a week earlier." By year's end I have amassed invaluable information to guide the next season. We used this practice for our first few growing seasons to create our seed starting plan, and we still use it when testing new crops.

An approximate English translation of *heijunka* is "load leveling." According to lean, when work is spread out and evenly paced, quality increases. When it is crammed and rushed, quality suffers. A key objective of our *Heijunka* Calendar is to help us level stress points. One of the stress points for us in the early years of our farm was midsummer. Spring planting had worn us out and yet there was no time for a break. Our mistakes escalated. Now we plan for a vacation every July. We stop our CSA and wholesale deliveries for a few weeks and limit what we offer at the farmers' market to an amount that an intern and perhaps a hired helper can handle.

Another goal of the *Heijunka* Calendar is to simplify planning (see figure 1.1 on page 6). With the calendar, our entire year fits onto one or two pages. We post the calendar in the processing area and in the seedling greenhouse for ready access.

The calendar is arranged weekly. Crops can be planted anytime during the week. Here, *DS* means crops are direct-seeded. Otherwise, crops are seeded into flats or paper pot chains in the greenhouse for transplanting to the field. To alleviate stress points, we shift seeding dates around—for example, seeding more in February and less in May.

How do we determine the number of seeds to plant? We refer to the *Kanban* Maps (see figure 1.2 on page 12) to determine the number of beds we want of the crop. Then we calculate the number of plants per bed. For example, let's say it is the first week of February. Our *Heijunka* Calendar tells us it is time to seed kale for an unheated greenhouse. Our *Kanban* Map shows that we want three beds of kale. So we will plant the proper number of seeds into a plug flat (or a bit more to cover possible seedling losses). The system is simple enough for anyone to follow.

Our basic farm plan is to start tomatoes for heated greenhouses in the last week of January. A week later we seed quick crops for interplanting between the tomato rows—head lettuce, baby greens, romaine, beets, turnips, radishes, and green onions—all as transplants. This gives us a full selection of food by late March—when most farms are just gearing up—and takes advantage of precious heated greenhouse space. After we harvest the quick crops, the tomatoes fill in the open spaces. Fall planting also begins early—the last week of May with storage carrots—and continues June through September with kale, beets, turnips, radishes, and greens. This long season gives us a wide selection of fresh crops going into late fall, when other farms are packing up for the year. We call this farming the shoulders because it focuses our production on the shoulder seasons when demand is highest—and gives us that July vacation.

A Plan for a Beginning Farmer

In the first year don't tackle every crop in the seed catalog. Pick just a few. Starting out, I would seed mixed salad greens and a hardy spinach variety, such as Gazelle, as soon as the ground is workable, and every week until temperatures linger consistently above 72°F to 75°F. Greens are a sure sell in most markets. Then I would seed determinate red tomatoes in time to set out after the last frost in your area, or a few weeks earlier if you have a greenhouse or tunnel. BHN 589 is a de-pendable variety. In the fall, when nights start to cool to below 68°F, I would seed greens again. If you can master greens and tomatoes in year one, you are off to a good start. Other easy crops for the first year include radish, sugar snap peas, head lettuce, romaine, kale, and green beans. If you want to try starting a CSA, take on 10 customers and keep your season to 12 weeks. Keeping it simple will more likely result in success, and that will build confidence.

Month, Week	Crops for Heated Greenhouses (1 and 2)	Crops for Unheated Greenhouses (3 and 4)	Crops for Field Plots
Jan: Week 1	Salad greens*	Salad greens	
Jan: Week 3	Salad greens	Salad greens	
Jan: Week 4	Tomatoes		
	Peppers		
Feb: Week 1	Salad greens	Kale	Shallots
	Quick crops**	Chard	Leeks
Feb: Week 3	Salad greens	Salad greens	
	Quick crops	Quick crops	
	Tomatoes	Tomatoes	
	Peppers	Pink ginger	
		Beets	
Mar: Week 1	Salad greens	Salad greens *DS*	Carrots *DS*
		Quick crops	Kale
Mar: Week 3		Salad greens *DS*	Salad greens *DS*
		Quick crops	Quick crops
		Tomatoes	Sugar snap peas
		Peppers	
		Beets	
		Green beans	
Apr: Week 1		Salad greens *DS*	Salad greens *DS*
		Quick crops	Quick crops
			Carrots *DS*
			Kohlrabi
			Bulb fennel
Apr: Week 3		Tomatoes	Salad greens *DS*
		Eggplant	Quick crops
			New potatoes *DS*
			Parsley
			Cilantro
			Green beans
			Cherry tomatoes

Figure 1.1. *Heijunka* Calendar Example: Weekly Seeding Plan

Note: All crops are grown as transplants unless marked direct-seeded (*DS*).

Month, Week	Crops for Heated Greenhouses (1 and 2)	Crops for Unheated Greenhouses (3 and 4)	Crops for Field Plots
May: Week 1		Basil	Salad greens *DS*
			Quick crops
			Edamame
May: Week 3			Salad greens *DS*
			Quick crops
			Storage carrots *DS*
			Zucchini
			Cucumber
June: Week 1			New potatoes *DS*
			Bulb fennel
June: Week 3		Kale	
July: Week 1			Carrots *DS*
			Beets
			Sugar snap peas
			Zucchini
			Green beans
July: Week 3		Quick crops	Salad greens *DS*
		Parsley	Quick crops
			Beets
Aug: Week 1		Quick crops	Salad greens *DS*
			Quick crops
			Carrots *DS*
Aug: Week 3		Quick crops	Salad greens *DS*
			Quick crops
Sept: Week 1	Salad greens *DS*	Salad greens *DS*	
Sept: Week 3	Salad greens *DS*	Salad greens *DS*	Garlic *DS*
Oct: Week 1	Salad greens	Salad greens	
Oct: Week 3	Salad greens	Salad greens	
Nov: Week 1	Salad greens		
Nov: Week 3	Salad greens		
Dec: Week 1	Salad greens		
Dec: Week 3	Salad greens	Carrots *DS*	

* Salad greens = baby lettuce, baby Asian greens, arugula, spinach.
** Quick crops = radish, Hakurei turnips, green onions, bok choy, multileaf head lettuce.

In this March greenhouse, we grow quick crops in the aisles between tomatoes, maximizing space and allowing for a fully stocked booth at a time of year when most farms are just gearing up.

Workers hold turnips and radishes seeded in January, transplanted in February, and then harvested March 25.

After quick crops are removed, we often lay tarps between the tomato rows. Soon the tomato plants will fill up the greenhouse.

Kanban Maps

Next we create maps that tell us how many beds of each crop we plan to grow and approximately where to locate them. Except for greenhouse tomatoes and peppers, we don't generally plan precisely where crops will go. Rather, we put them in where there is an opening.

Kanban translates to "replacement signal." For instance, in the days when milk came in glass bottles, an empty milk bottle sitting in front of a house was a *kanban*—a cue to replace. In lean factories, *kanbans* ensure smooth production flow. For example, a parts bin might include a *kanban* tag; when the bin is empty, workers turn in the tag to the stocking department so that the bin is quickly replenished.

Open beds on our farm constitute *kanbans*. We fill them as soon as we can—often within hours—keeping the farm at full capacity. We always try to have a few flats or paper pot chains of transplants ready for these openings. With few exceptions, all greenhouse beds are full all year and all outdoor beds are full April through November. For soil health and to mitigate disease, we typically don't replace a crop with another of the same family, and, as I will explain more later on, we direct-seed into our most weed-free beds. Beyond that, little planning is required.

Clay Bottom Farm. Our *Kanban* Maps correspond to spaces in this illustration.

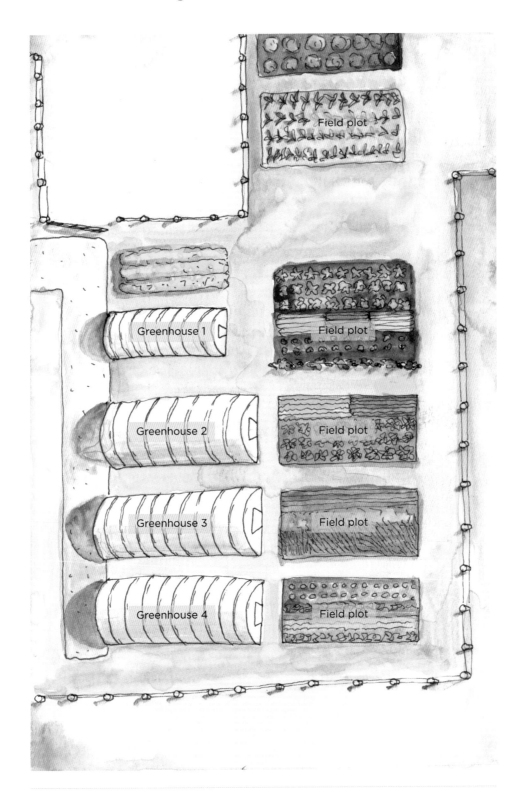

Field plot

Greenhouse 1

Field plot

Greenhouse 2

Field plot

Greenhouse 3

Field plot

Greenhouse 4

Field plot

Crops for Greenhouse 1

4 BEDS

Bed size: 24″ × 50′ — Heater turned up to 55°F on March 1

| BHN 589 tomatoes (35) |
| Heritage tomatoes (35) |
| BHN 589 tomatoes (35) |
| BHN 589 tomatoes (35) |

Totals for Greenhouse 1 and 2

Carolina Gold tomatoes:	50
BHN 589 tomatoes:	105
BHN 871 tomatoes:	50
Early Sunsation peppers:	50
Glacier tomatoes:	25
Heritage tomatoes:	235
Red Knight peppers:	100

Crops for Greenhouse 2

10 BEDS

Bed size: 24″ × 42′ — Heater turned up to 55°F on April 1

Early Sunsation peppers (50)	Glacier tomatoes (25)
Heritage tomatoes (50)	BHN 871 tomatoes (50)
Heritage tomatoes (50)	Carolina Gold tomatoes (50)
Heritage tomatoes (50)	Heritage tomatoes (50)
Red Knight peppers (50)	Red Knight peppers (50)

When tomatoes finish, all heaters are turned down to 28°F for winter crops. Salad greens* and quick crops** planted between tomato and pepper rows.

Crops for Greenhouse 3

16 BEDS

Bed size: 30″ × 42′ — Unheated

Crop	No. of Beds
Salad greens*	4
Quick crops**	3
Tomatoes	3
Basil	2
Carrots	1
Chard	1
Eggplant	1
Peppers	1

Crops for Greenhouse 4

12 BEDS

Bed size: 30″ × 42′ — Unheated

Crop	No. of Beds
Beets	3
Green beans	3
Kale	3
Pink ginger	2
Parsley	1

Note: Except for tomatoes and peppers (Greenhouses 1 and 2), the *Kanban* Maps don't indicate placement. They just dictate number of beds. Crops are planted in any open bed.

*Salad greens = baby lettuce, baby Asian greens, arugula, spinach.
**Quick crops = radish, Hakurei turnips, green onions, bok choy, multileaf head lettuce.

Figure 1.2. *Kanban* Map Example: Greenhouses Spring 2017

Crops for Field 1 5 BEDS	No. of Beds SPRING	No. of Beds FALL
Shallots	2	
New potatoes	1	1
Salad greens*	1	4
Zucchini	1	

Crops for Field 2 9 BEDS	No. of Beds SPRING	No. of Beds FALL
Salad greens*	1	1
Quick crops**		1
Kale	3	3
Carrots	5	4

Note: Except for tomatoes and peppers (Greenhouses 1 and 2), the *Kanban* Maps don't indicate placement. They just dictate number of beds. Crops are planted in any open bed.

 * Salad greens = baby lettuce, baby Asian greens, arugula, spinach.
 ** Quick crops = radish, Hakurei turnips, green onions, bok choy, multileaf head lettuce.

Crops for Field 3 34 BEDS	No. of Beds SPRING	No. of Beds FALL
Carrots	3	8
Beets, red	3	3
Beets, golden	2	2
Quick crops	2	5
Salad greens	3	5
Sugar snap peas	1	1
Beans	1	2
Edamame	1	1
Bulb fennel	1	1
Leeks	2	2
Kohlrabi	½	
Cilantro	½	
Onion plants	1	
Parsley	½	½
New potatoes	2	1
Garlic	8	
Chard	½	½
Cucumbers	1	1
Cherry tomatoes	1	1

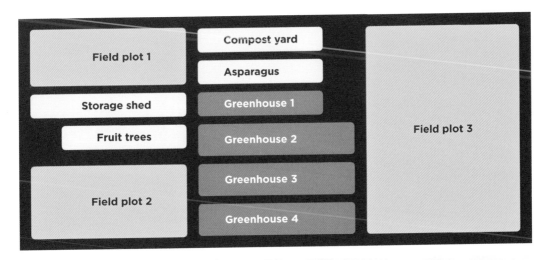

Figure 1.3. *Kanban* Map Example: Field Plots Spring 2017

Choosing Seed Varieties

When choosing seed varieties, look for clues in seed catalogs that tell you the seed is popular with market growers. Those crops perform best. Also, don't get too stuck on your favorite varieties. With constantly improving seed genetics, stay on top of trends. We use a 15 percent rule: 15 percent of what we grow should be new every year, the rest familiar. More than 15 percent new items and we are probably experimenting too much. Less than that and we will miss out on innovative seed developments. Appendix 2 lists our current varieties.

With *kanban* planning we replace beds as soon as they open up rather than relying on complex multiyear maps. This simplifies planning and ensures we operate at full capacity. The only crops for which we determine the precise location and number to grow are tomatoes and peppers grown in heated greenhouses.

We create one map for greenhouse crops (figure 1.2) and one for field crops (figure 1.3). We keep separate maps for spring, fall, and winter. We determine how many beds to set aside for each crop based on orders and projections. Amounts change all the time. For example, in 2015 we sold out of fall carrots at our farmers' market by late October; in 2016 we increased the number of winter harvest carrot beds from six to ten.

We don't rigidly hew to the maps. They are merely rough guides. When beds open up, we consult the maps but also check our plans against current reality: what our customers most want *right now* takes precedence. For example, in 2015 a bed in which we had planned to seed arugula—a crop with a lower profit margin—opened up late July. Just a week before, a chef told me he would like fall bulb fennel for a new fennel slaw dish on his menu. I ditched the arugula and planted the fennel. Success as a niche market grower requires this type of nimble planning. The point in treating beds as *kanbans* is to account for all beds and to make sure they are always filled with the most valuable crops.

Lean Pull: Collecting Orders First

When you produce first and then hunt for customers, the sequence is called *push* production. Instead, lean advocates *pull* production: lining up sales in advance, and letting customers *pull* your farm along. It is all right to farm with a mix of pushing and pulling, especially in the beginning stages when you need to work out production kinks and might not want to promise too much. But pull systems are more efficient because they lessen the possibility that you will overproduce.

At first, while we built up our business, our planning relied mostly on hunches as to what would sell. We suffered a lot of overproduction as a result. Now we secure as many advance orders as we can—either specific orders or projections based on past sales. Our goal is to decide very little on our own regarding what to grow. Here are three practices we use to generate orders first.

1. Selling ahead through a CSA
2. Generating wholesale orders
3. Tracking farmers' market sales

1. SELLING AHEAD THROUGH A CSA

With a CSA, customers pay up-front for an agreed-upon amount of food delivered once per week. I know of no better way to pre-sell food: each sign-up equals one giant order.

The roots of the CSA model can be traced, in part, to black history. Professor Booker T. Whatley, of Tuskegee University in Alabama, was considered a twentieth-century pioneer in sustainable agriculture. In the 1970s he even toured the country promoting "smaller and smarter" farming through what he called "Clientele Membership Clubs." The model involved households—"city folks, mostly"—paying in advance to receive food directly from a farmer. He said, "This enables the farmer to plan production, anticipate demand, and of course, have a guaranteed market." His goal was to regenerate an agrarian black middle class.[1] A version of the CSA model, called Teikei, was also used by Japanese farmers in the 1960s specifically to combat waste stemming from overproduction. We solicit CSA customers through email, Facebook, and flyers. Customers sign up through our website, where they pay and choose their box size and pickup locations.

According to many farmers I have talked with and to recent data, the CSA model in many places is under duress. One survey conducted by Small Farm Central found that the average CSA farm in 2015 lost most of their customers by 2016: the customer retention rate was just 46.1 percent.[2] One reason many current CSA farms struggle to keep customers, I suppose, is that the novelty has worn off. Whenever a new service arrives, customers

A Clay Bottom Farm CSA box. Notice the four corners: greens (kale and romaine), tomatoes, an allium (leeks and onion), and a fresh snack (French Breakfast radishes). Customers tell us that they want these four groups of food every week. They are the "legs" of our CSA table, and we supply them for as long a season as we are able to. In addition, we add seasonal items. This week it's potatoes and parsley.

How Much Money Can You Make?

An acre of land contains roughly 43,000 square feet. If you dedicate 13,000 square feet—around one-third of it—to pathways, that leaves 30,000 square feet on which to produce crops. In most areas of the United States, at least two crops can be grown per season on a spot of land. Take head lettuce as an example. A relatively simple-to-grow crop, head lettuce occupies about a square foot of ground, and might sell for around $2 per head. One crop of head lettuce, then, could yield $60,000 per acre; two crops could bring in $120,000. Higher-dollar-value crops—such as heirloom tomatoes—could in theory gross at least twice that.

The math seems simple, but of course reality is more complicated. Who is to say that your customers want 30,000 heads of lettuce, all at one time? Most market farms will need more diversity than that, which adds complexity and costs. More fundamentally, the 10 wastes I outlined in the introduction—from overproduction waste to defect losses to wasted motion—forever gnaw away at profits.

Our vision has been to keep the math simple. We hope to earn a maximum yield from every square foot by cutting out the 10 wastes. Achieving low-waste production does require investment. Beyond the purchase of our land, we have invested in greenhouses, a processing area, a cooler, and other pieces of infrastructure (see chapter 12), plus specialty equipment and hand tools, to help us reduce wastes.

In appendix 3 I list the dollar value a bed 30 inches by 100 feet can yield in our markets. The list shows that yield per bed can range widely, and certainly not all crops reach an optimum yield every season. Still, the list might help you project gross income as you plan your market farm.

are willing to put up with inconveniences to try it out. In the early days of CSAs, customers sometimes drove long distances, often at inconvenient times, to participate in the novel system. With CSAs now common, enthusiasm has waned. Also, mid-level grocers now carry organic food, as do many big-box stores, offering organic celery, Band-Aids, and blue jeans in one stop. CSA farms are no longer a unique option for buyers of natural food.

All of this puts pressure us to step up service. Several years ago we added refrigeration units to our pickup locations, allowing customers to pick up whenever it suited them rather than during a short window of time. More recently we have partnered with our local food co-op, where many of our customers shop, to serve as a pickup location. Customers pick up their CSA produce when they do their other shopping, saving a trip. We also run CSA "tabs" at our farmers' market booth. Customers pay us up-front, as with a standard CSA, but instead of picking up a box every week, they come

Lean and People on the Farm

It is one thing to practice lean ideas on your own, as one person, but it is another to practice lean with other people on the farm. In part that is because lean is more than a set of skills; lean is a set of ideas. It is relatively easy to teach skills, but more difficult to transmit ideas. Still, here are my lean tips for working with staff:

1. *Practice heijunka (load leveling) with people.* When a workload is uneven, lean managers try to level it out, decreasing stress in the workplace. A good practice is to check in regularly with workers to assess whether they are working too many hours. Burnout prevention is a skill we are still learning for ourselves as well as others, and it is essential for long-term farming. Instead of keeping a rigid schedule, we often call off early (especially on those hot afternoons), or take a day off entirely to recharge. If a worker seems to be dragging, I find another person for the job. Emotional *heijunka* keeps our farm moving ahead with energy.[3]
2. *Cut out the* muda. This means cut to the chase. It is easy to skirt around problems and let disagreements fester. This just wastes time. A better approach is to quickly address problems. Don't be rude, but be respectfully direct.

   For instance, if I see poor performance, I will calmly point out what is not being done correctly as soon as I see it, show a better way, and then allow time for correction. I have discovered that it is important to explain not just *how*, but also *why* a task should be completed in a certain way. The more natural I am, the better. Say a worker cuts head lettuce too high, causing loss of some lower leaves. With a friendly demeanor I might say, "Hey those heads look great, but that's not quite right." Then I will cut a couple of heads to demonstrate, explain why a lower cuts works better, and then give the worker space to improve. I have found workers actually appreciate a direct-though-friendly approach. They want to improve as much as we do.
3. *Practice ruthless* seiketsu *(standardizing).* The fourth step in lean's 5S organization system is *seiketsu*, or

to our booth anytime it's open and choose whatever items they want. We keep track of their balance in a notebook. If they pick up just one tomato or a mix of 10 items—or if they skip for an entire month—that is their prerogative. We give them a discount for paying up-front, and if they want large quantities—say a bushel of tomatoes—we ask for a bit of advance warning.

With all this flexibility, how do we decide what to grow? While we can't plan with complete precision, we can use surveys to learn to as much as possible what CSA customers want. Twice each year I email a list of questions:

standardizing (for more on 5S, see *The Lean Farm*). In this context *seiketsu* means completing tasks the same way every time. This is not just for your sake, but also for your workers'. It is unfair to them to rely on haphazard systems, understood by you but no one else. Whenever I develop a new system, whether with the paper pot transplanter (see chapter 6) or in the processing shed, my goal is to design the task so that *anyone* can do it. The reward is that new workers start building value from day one.

The lean system is much more powerful when employed by the whole team.

1. Have we offered the right mix of foods for you?
2. What else might make it easier for you to receive your food?
3. Has the amount been too much, too little, or just right for you or your family?

The questions address *what*, *when*, and *the right amount*. We will never manage to please everyone, of course, but the answers we receive give insights useful for planning. The key in surveying customers is to keep it simple. Long-form questionnaires don't get answered.

2. GENERATING WHOLESALE ORDERS

We make a point of talking to each wholesale buyer in the winter, assessing what we might grow more or less of the following year. We don't expect each buyer to follow through with every order discussed, but the conversations give us information with which to plan. Chefs appreciate this. To please chefs, we sometimes grow less common crops like Thai hot peppers for them, even when these crops represent minor profits for us. They keep long-term accounts coming back and keep our work interesting.

In addition to communication with chefs, we run a QuickBooks report at the end of the year to tally sales of each item per wholesale customer. This helps us track best-sellers.

The best way to create pull is to discover precisely what customers want, when they want it, and in what amount. Here I discuss orders with a chef.

3. TRACKING FARMERS' MARKET SALES

Of course, farmers' market customers don't give us an advance order at the beginning of the year. To predict how much to send to a farmers' market, we track sales from the previous week's market. We do know of a few annual trends—the Fourth of July market will see a bump in sales, for example—but for the most part we send to market the same amount that sold at the previous market, or a bit more.

To track sales efficiently, we use a magnetic whiteboard hung in our processing area. By Monday morning, the day we harvest for a Tuesday market, I write a number next to each vegetable in the "Farmers' Market Tuesday" column that communicates to pickers how much to harvest—for example, 40 bunches of curly kale or 30 bunches of carrots. After market, I ask the person who staffs the booth to send me a text with a list of what is left over—say 20 bunches of curly kale or 2 bunches of carrots. If an item sold out, the list includes the time. Say beets sold out at 10:30 a.m., midway

Our magnetic whiteboard, posted at the center of our processing area, documents weekly orders. Bicolor magnets indicate progress: green means an item is picked and is in the processing area; red means it's packaged and customer-ready in our cooler. The board is a version of *kamishibai*, a lean visual organization method. Photo courtesy of David Johnson Photography.

through our market. This tells me we should take twice as much to the market next week.

When I see items consistently selling out or coming home unsold, I jot it down in a notebook. We use these notes when we create the following year's *Heijunka* Calendar and *Kanban* Map. For example, we noted in 2016 that kale sales dropped off in mid-May, when lots of carrots and turnips became available. In 2017 we planned for less kale at that time of year. Practices like these keep our farming tightly aligned with what customers want.

CHAPTER 2

Leaning Up
Bed Preparation

*Is it practical to run the garden exclusively with the
use of compost, without the aid of the so-called chemical or
artificial fertilizers? The answer is not only yes, but in such case
you will have the finest vegetables obtainable, vegetables
fit to grace the table of the most exacting gourmet.*

—J. I. RODALE

We first toured our current farm as prospective buyers in early September 2008. It was a calm evening, with a slight chill in the air. The evening sky was pink and orange, and corn and soybean crops all around us were drying down and turning brown, nearing harvest. The field behind the barn was full of soybeans. We thought it would make a fine spot for vegetables.

I pulled out a shovel that I had brought along and dug a hole, pushing hard, between two soybean rows. It was pure clay, packed down from years of being driven over by tractors and combines.

A few days later we talked with some locals about the "Millersburg mud sock," as the area surrounding our farm is somewhat derisively called. Was it possible to grow vegetables in such heavy clay? My shovel test had me worried.

They told us that even though they set their vegetables out late every year, waiting for the soil to dry out, their plants always caught up. In fact, because clay holds nutrients so well, crops usually out-yielded those grown in sandier ground. In addition, growers rarely watered the ground, because it held moisture so well. The trick, said one neighbor, was to "stay off it while it's wet." We drove around to see if all this was true. Sure enough, almost every farm boasted a healthy vegetable patch.

Convinced, we purchased the farm.

In retrospect, although our clayey soil has served us well, I would suggest that new farmers start with a sandy loam soil, if possible. Clay took a few seasons of focused work to loosen and build up. Here is what we did. That first fall, once the soybeans were gone, we hired a neighbor with a large tractor to chisel-plow our growing area, loosening up the years of compaction. Then, with our Ford 8N tractor I disked, leveled the soil with a cultipacker, and seeded a perennial cover crop, a mix of orchard grass and clovers, over the entire field. The next spring we marked out growing plots and plowed the grass and clover in, leaving the cover crop to grow in the aisles.

In addition, we tested soils all over the farm and applied bonemeal, alfalfa meal, and other amendments to balance deficiencies. We seeded sorghum sudangrass, crimson clover, and vetch as cover crops.

Even after the first year, we continued this approach—test, amend, and plant cover crops—to bust up the packed clay and to keep the soil healthy and alive. While this was indeed improving the land and growing ever-healthier crops, progress was slow and costly. The tests, cover crops, and copious amendments cost us precious time to manage. More than once we missed an important harvest, or cut corners with trellising tomatoes, because we were busy amending soil. Also, cover crops occupied valuable space for months on end. On our small plot, they limited how much we could grow. They also cost us a lot of money, as we poured hundreds of dollars every season into the soil. We started to wonder whether, given our tiny size, we could build our soils faster, at lower cost.

After a few seasons we developed a new approach. We replaced cover crops, fish emulsion, and most other fertility sources with compost. (For tomatoes, we sometimes amend the compost with a small amount of minerals if a soil and leaf tissue test indicates the need.) With the right equipment, our compost is quick to make and low effort to apply. The system uses resources—manures, grass clippings, leaves, and more—that would otherwise be wasted in our local community. Because these are free or low cost for us to use, composting is cheaper than other fertility methods. After a period of years of being amended with compost, composted soil starts to look and feel like chocolate cake. It is a pure pleasure to work. The most important benefit, however, is crop quality. We can plant almost anything anywhere on the farm and expect resilient, nutrient-dense, highly marketable crops.

Our bed preparation now consists of four steps:

1. *Amend* the soil by adding compost, usually based on observation of soil and plants, and occasionally based on soil tests.

2. *Aerate* with broadforks, a chisel plow, or our root digger.
3. *Blend* only as absolutely needed to incorporate compost or crop residues with a claw hoe, walk-behind tractor, or rotavator implement.
4. *Shape* with shovels and rakes or a bed-shaper implement, or by simply walking up and down aisles, compressing the pathways.

After shaping, we rake smooth with a 30-inch bed rake if needed, and then seed or transplant. While this method works well in clay, we would use it even on sandier ground, perhaps with less aerating. In *The Lean Farm* I explained this system in broad strokes. Here it is in detail.

Step 1. Amend with Compost

We keep two types of compost, "brown" (high-nitrogen) compost and "black" (low-nitrogen) compost.

1. *Brown (high-N) compost.* Our "brown" compost, composed exclusively of animal manures and bedding, is dark brown when finished. It is a high-nitrogen (N) mix, with plenty of phosphorus (P), potassium (K),

We keep two types of compost: a high-N "brown" compost (*background*), and low-N "black" mix.

and magnesium as well. We use it primarily as a fertilizer boost rather than to build soil structure.

Indiana is the largest duck-producing state in the nation, so most of our manures come from an organic duck farm that uses sawdust for bedding. We pay extra for an organic source to avoid possible contamination in the manure or bedding. If manure for compost does not come from an organic farm, be sure to investigate the source of bedding anytime you purchase manure, since many farms now use shredded chipboard, which is full of glues, chemicals, and sometimes paint. We have composted manures with straw and with sawdust bedding, and as long as we compost them properly, both perform well.

2. *Black (low-N) compost.* Our "black" compost is composed mostly of "green" manures: grass clippings, municipal leaf mold, moldy hay, straw, and whatever other plant-based organic matter we can find. This high-carbon mix is lower in N, but full of micronutrients such as boron, calcium, magnesium, phosphorus, and potassium. We use it heavily on fruiting crops (where too much nitrogen would lead to excessive foliar growth) and to build soil structure.

For a complete description of how we make black and brown composts, see chapter 3.

At first, our soils showed uniformly poor health. Tests of our soil 6 inches down revealed low levels of P and organic N, and organic matter was less than 1 percent. Soils exhibited excessive water retention and crusting. On beds where we had added several inches of compost, conditions improved remarkably fast. Nutrient amounts rose to levels required by our

Skim-coating garlic beds with high-N brown compost. We will blend this in shallowly. Compost is our primary source of fertility.

crops, and organic matter rose to more than 10 percent, even 6 inches deep. Hard, crusty ground turned into soft, moderately drained soils. We realized that with compost we could achieve great soil anywhere on our farm.

The process was not overnight. To bring our soils up to par, across our 1-acre plot we added up to 2 to 3 inches of black compost and ½ to 1 inch of brown compost for a few growing seasons, totaling 8 inches or more in some places. We blended our composts into our native soil to the depth of our tiller, about 6 to 8 inches. All soils are unique, and yours might require more or probably less. To determine what type and how much compost to add to a brand-new plot, I recommend starting with soil and compost tests from a laboratory, then adding composts a few inches at a time until you notice an increase in soil workability and nutrient deficiencies lessen.

In the following years we dialed back applications of compost because our soils retained residual nutrients from the heavier applications early on. Also, after the initial buildup, we apply compost only where it's needed—in transplant holes or spread across a bed surface—rather than uniformly across an entire plot. We periodically test soils to ensure P and N levels don't accumulate to excessive levels. With the exception of tomatoes, however, we don't pull soil tests prior to growing every crop. That would require more time than we have available.

Instead, when deciding how much compost to use year-to-year, we supplement occasional soil testing with *observation*. First, we observe the soil: Is the soil fluffy enough to scoop up by hand, or is it crusted over? Does it look alive or inert? Can we see or feel humus, which is dark and crumbly, or does it feel like pure clay or sand? Does the soil drain slowly or quickly? Both might be cause to add compost, which evens water retention and

Compost for Soil Building versus Short-Term Nutrient Release

I recommend focusing on higher-carbon (low-N) composts for initial soil building. Our black compost contains a high C:N (carbon-to-nitrogen) ratio. With its long nutrient release, it is invaluable for building up soil. On the other hand, our high-N compost, which costs us less, performs best at supplying short-term nutrient needs.

A major of advantage of using compost is that nutrients from compost are held in a relatively stable form and are released slowly in the soil over a period of years, especially when compost with a higher C:N ratio is used. Compost pays you back long-term.

Compost: How Much Is Too Much?

While in our experience there is no better fertilizer than compost from your own farm, there is such a thing as too much compost. In particular, when composts made from animal manures and bedding, which contain relatively high levels of P and N, are applied to soils already high in P and N, excessive amounts quickly accumulate, resulting in nitrate leaching, phosphorus runoff, accelerated eutrophication (a form of pollution) of waterways from excess nitrogen, and consequent excessive vegetative growth. Some states regulate P levels in soil to reduce pollution. For example, Vermont requires farms with soil test P levels above 20 ppm to have a plan for reducing P applications. Soil and composts tests will reveal these amounts.

For a detailed discussion on the specific hazards of too much P and N, see the New England Vegetable and Fruit Conference's paper, "Soil Organic Amendments: How Much Is Enough" (see "Composting Resources" on page 222).

draining. Judging these conditions well takes experience, but over time we have developed quick reflexes.

Second, we observe plants, which we categorize as either generative or vegetative.

1. *Generative crops.* These are crops that yield fruits, not leaves. Examples include peppers, tomatoes, watermelon, and eggplant. We observe them for their fruit-to-leaf ratio. Ideally, fruits are large and plentiful and leaf cover is sufficient to shade fruits and weeds, but not overwhelming. Regarding eggplant, for example, an oversize bushy plant with dark green leaves yet small eggplants often indicates too much N in the soil. Our yields will be low. We grew too much plant and not enough fruit.

   When transplanting generative crops, we usually add more "black" (low-N) than "brown" (high-N) compost. If fruiting seems weak and leaves are abundant, we cut down on the brown compost next time around. On the other hand, if the plant overall seems weak with slow growth, we add more brown compost.

2. *Vegetative crops.* Vegetative crops are plants that we steer toward leafy growth, such as kale, spinach, head lettuce, and other greens. We also include root crops and alliums in this category; even though we don't harvest their leaves, we have found that they appreciate a similar treatment. Are leaves dark green or have they turned pale green or even yellow? Are shallots, beets, and potatoes growing to full size? If

Crop steering with compost. We apply a side dressing of mostly low-N black compost (with added organic amendments) to tomatoes once per month. Side-dressing shallots with high-N brown compost.

not, why not? To address such problems, we might add a higher ratio of brown compost at planting time to boost leafy growth, or apply a side dressing midseason. A typical formula might be 3 parts brown to 1 part black.

Third, we take note of a previous crop. For example, if a heavy feeder such as kale follows spinach, another heavy feeder, we double the amount of brown compost, since spinach leaves little nitrogen behind. If radish, a light feeder, follows spinach, we usually add nothing, since there is likely enough nitrogen already in the soil. If salad greens or bok choy follow tomatoes, we add nothing, since these crops are light feeders. Sufficient nutrients are left in the soil after the tomatoes leave. When we are unsure about our observation, we might pull a soil test to confirm.

There is no set formula, because every crop that goes in has unique nutrient needs, and every crop coming out has altered the soil in some way. Every year we learn something new about what to expect when kale follows potatoes, or greens follow tomatoes. Even so it is possible to provide a rough guide to this observation-based method (see table 2.1 on page 31). For additional resources on small-farm composting, see also "Composting Resources" on page 222.

Moving Compost

To move compost about the farm, we at first relied on scoop shovels and garden carts. Once we purchased our John Deere Gator, we began to use it for all of our compost hauling. Now we happily own a skid loader that scoops, moves, and dumps hundreds of pounds of compost with only a few hand motions.

If we are adding compost at the beginning or end of a season to an entire plot, we wait for dry weather in order to minimize compaction and then drive the loader directly onto the plot. Usually, however, we keep the loader out of the plots, instead dumping compost on the ends of beds or just inside the door of a greenhouse, and moving it to the middle with shovels. This is one reason we appreciate short beds and wide greenhouse openings.

A skid loader or tractor with bucket, combined with wide greenhouse openings, takes effort out of moving compost. Our loader is a Gehl 4460 with 44-horsepower (hp) diesel engine. Skid loaders allow for zero-radius, or "pirouette," turning, an advantage on a small farm.

Traditions like permaculture and Korean natural farming, among many others, espouse other ideas for amending soil. Examples include fermenting sugars and plant juices in glass or ceramic containers and adding them to the soil, using fish amino acids and oils, and extracting and fermenting maltose, made from sprouted barley, for use as an amendment. Happily, there is no one right way to grow food. On our farm, we have used various fish blends, molasses, and sea products, we have collected worm castings and made compost teas, and we continue to experiment.

To be clear, while we no longer rely heavily on cover crops and

Deeply composted soil is a pleasure to work.

Table 2.1. Soil Amending for Specific Crops at Clay Bottom Farm

Generative Crops	How We Fertilize
Cucumbers	1:1 ratio of brown to black compost, applied 1 inch thick across bed surface at beginning of season.
Eggplant	1:4 ratio of brown to black compost, 1 inch thick.
Green beans	Usually no amending needed.
Peppers	1:2 ratio of brown to black compost, 1 inch thick.
Summer squash	Usually no amending needed.
Tomatoes	1:4 ratio of brown (high-N) to black (low-N) compost, about 1 quart per plant once a month. Granular amendments sometimes added to compost.
Vegetative Crops: Light Feeders	
Beets	1:1 ratio of brown to black compost, ½ inch thick.
Carrots	Usually no amending needed.
Radishes	Usually no amending needed.
Salad greens	Usually no amending needed.
Turnips	⅛ inch of brown compost.
Head lettuce and romaine	Usually no amending needed.
Vegetative Crops: Heavy Feeders	
Alliums: garlic, shallots, leeks, onions, spring onions	1 inch of brown compost mixed in a planting time.
Basil	For robust basil, ½ to 1 inch of brown compost, though we keep an eye out for aphids, indicating too much N.
Kale	For full-size plants, 1 inch of brown compost. For baby leaf, usually no amending needed.
Pink ginger	1 inch of brown compost mixed in at planting time.
Rhubarb	2 quarts of brown compost, applied every fall to each plant.
Spinach	¼ inch of brown compost.

Note: These rates are specific to our own soil and reflect reapplication rates after initial soil buildup. Develop your own plan based on conditions on your farm. We spread composts across the bed surface, not in aisles. We shallowly till in composts before seeding or transplanting.

minerals, we are not opposed to their use. Amending with the right mix of minerals can bring an imbalanced soil to life. And cover crops, for those with extra room and time, will release nutrients into the soil, improve texture, and hold open ground from runoff. I do not cover these systems in detail in this chapter because we rarely employ them, and because many other resources cover the topics in depth. I discuss a lean approach to cover crops for weed management in chapter 8.

Step 2. Aerate

Deeply aerating soil, especially in heavier-texture soils, augments the work of compost. Simply put, vegetable roots want loose soil. Aerating allows oxygen to reach those roots, and softens ground so they can penetrate deep and grow uniformly. Also, we have found that tillers work more effectively on pre-aerated ground.

We use three tools to aerate. First, we use digging forks to loosen up ground in the tight corners of the greenhouse or if we are transitioning a small part of a bed from one crop to another. We step on the fork until it is fully in the ground and tilt it toward us, working backward so we are not stepping onto loosened ground.

Second, we use a broadfork, a two-handled tool with 10-inch-long tines designed specifically for the purpose of aerating soil. The key to efficient use is to develop a rhythm: hop onto the bar, pull the handles back to lift the soil, hop off, and repeat. Our rule: the bar holding the tines should touch the ground before the tool is pulled back. This ensures maximum aeration. There is no need to completely flip the ground. Just lifting it

Three ways to aerate soil: broadfork, four-shank chisel plow, and root digger. We installed a quick hitch system from Jiffy Hitch, seen on the root digger, on all of our implements so that we can swap them out quickly and safely.

slightly will do, and requires much less effort. We move the broadfork about 10 inches after each dig.

Before we owned our tractor, the broadfork was our most-used hand tool. But as our production grew, aerating with the broadfork was beginning to wear us out. Broadforking in clay soil is much more difficult than in sandier ground. Now we perform most of our aeration with our Kubota L3400 compact tractor and a four-shank chisel plow. I call the chisel plow a "broadfork on steroids." Entire plots can be aerated in minutes and at a deeper depth than the broadfork can achieve. An alternative is to use a root-digging implement set as deep as possible, which also raises the soil and introduces air.

We aerate every late fall after we have tilled in crops for the year. I plow at a different angle every year, rather than straight on. This practice ensures that soils are evenly loosened over time. Also, to aerate as deeply as possible, we welded a rock box for weight on the front end of our tractor. Front-end weights accomplish the same task. As I plow I keep an eye on the front wheels. When they spin, it's time to slightly lift the plow. This late-fall chiseling opens up soils so that they are more porous when snows melt, allowing them to drain faster in the spring.

We also aerate during the season, between crop plantings. For example, when we remove tomatoes from the summer greenhouses, we irrigate the soil to loosen it (we keep soil fairly dry for late-season tomatoes) and bring the compact tractor and chisel plow into the greenhouse for a quick round of aerating. We use a broadfork in the corners where the tractor can't reach.

The tractor also works well to aerate smaller patches. For example, when a bed of long-season kale is finished, we might pull the plants and then chisel-plow the bed if it needs loosening, preparing it for the next crop. The tractor and chisel plow are the correct size for this task (see the "Tractor Buying Tips" sidebar on page 34) so that they straddle the bed. The amount of time and effort saved as compared with aerating with a broadfork is tremendous, considering that we "flip" dozens of beds each month in peak growing season.

Step 3. Blend

The third step is to blend compost and soil together. Blending ensures an even distribution of the compost and breaks up larger clumps. For soil building, we work compost in as deeply as our tillers will allow. To supply short-term nutrient needs, we often blend compost into just an inch or two of soil. Blending also softens the ground, making it easy to transplant into and to run seeders.

Tractor Buying Tips

In recent years compact tractors have proliferated as more and more hobby farmers look for ways to tidy up their properties. Implements such as backhoes, trenchers, and bed shapers are now also scaled for compact tractors, making the machines more versatile. They now come with standard safety features such as rollover bars and automatic shutoff switches. With smoother steering and improved balance, they are also easier than ever to operate.

In general, tractor companies offer three types: hobby or sub-compact tractors, often designed for lawn mowing and light-duty yard tasks; compact tractors for small farms and orchards; and full-size models for commodity crop farmers and the construction industry. For the small farm, most sub-compact tractors do not feature the horsepower needed for plowing and chisel-plowing. Full-size tractors are awkward to maneuver in greenhouses and on small patches. The best fit is the compact tractor, designed for professional use, which strikes a balance between size and power. Here are points to consider:

Power. I recommend at least 33 hp if you are farming sandy ground, or 36 hp on heavy ground. Our Kubota L3400 with "xTra Power" and a 36 hp rating pulls a four-shank chisel plow in heavy soil, our heaviest task. When we are at "full pull" I am grateful for every ounce of power. There is no reason to buy more hp than you need, however. High-horsepower tractors require larger and heavier frames. I recommend staying as small as you can.

4WD versus 2WD. Four-wheel drive, which greatly increases pulling power, is a standard feature on many but not all compact tractors. Make sure yours has it. Two-wheel-drive tractors haul wagons well, but for serious pulling four-wheel drive is a must.

Center-to-center tire spacing. Tractors come with varying tire spreads. Ideally, you can match the distance between tires with your bed width. We searched for a tractor with slightly more than 30 inches between the inner surfaces of the tires because most of our beds are 30 inches wide. Sometimes, in order to define beds, we simply drive the tractor up and down a plot, letting the tires mark the aisles.

Tire type. Many tractors come delivered with turf tires. These low-tread tires work well on lawns, preventing ruts, but they are not ideal for farming. They are too wide and lack the traction needed to pull implements well. Instead, I recommend "ag" tires. This is the name often used for tires that are deeply ribbed and skinny. Ours are 11 inches wide, perfect for a 12- or 18-inch aisle.

Turning radius. The turning radius on tractors varies widely. On a small farm, you want as tight a radius as possible. We allow just 12 feet as a tractor turnaround at the ends of our plots. Being limited for space shouldn't preclude owning a tractor.

Front-end weights. I recommend front-end weights to counterbalance heavy implements being pulled by the tractor. This allows for deeper plowing, and the better balance means the tractor is easier to operate.

Hydrostatic versus manual transmission. Hydrostatic transmissions increase or decrease speed smoothly and automatically, similar to automatic transmissions in cars. Manual transmissions require pressing the clutch to shift gears. Having used both, I prefer manually shifting gears because I like to stay in one gear, at one speed, for many tractor tasks. In addition, hydrostatic transmissions have much higher up-front and maintenance and repair costs. However, both systems can work for the small farm.

Diesel or gas. I prefer diesel. Although there are many good gas-powered tractor engines, diesel engines offer more power for their size, achieve better fuel economy, and are made of more durable components. On balance, they outperform and outlast gas counterparts.

Category 1 hitch. Hitches come in five classes, ranging from 0 to 4. Most small farm implements are designed to fit a category 1 hitch. It is possible to convert implements from one size to another, but it is costly to do so. I recommend a category 1 hitch when you order your tractor.

New versus used. In general, we are fans of saving money on used tools that are in good shape. In the case of our tractor, however, we couldn't find one that fit all of our needs for a significantly lower price than a new one. Compared with other pieces of equipment, tractors hold their value well. Also, financing is easily available on most new models. We bought ours with a zero-interest five-year loan.

Of course, tractors are not for everyone. A good tool fits both farm and farmer. Some prefer smaller tools, like walk-behind tractors, or farming with horses, as a personal choice. In some locations, such as on very hilly ground and in cities, tractors are not always practical. However, for many small farms, compact tractors are an increasingly affordable way to gain a giant boost in production with less work. Many tractors costing between $15,000 and $20,000 will meet a small farm's needs for many years.

When should you upgrade to a tractor? In my experience, in the first few years you don't need a tractor. On our farm we hired out tractor work to others for the first four seasons. Once you are established, however, a tractor purchase might quickly pay off, depending on the scale of your operation.

We use four blending tools. First is a three-tooth cultivator, also called a claw hoe, which we use in areas smaller than about 100 square feet. I recommend a long-handled version of this tool to save your back. Second, also for small areas, we use a Honda mini-tiller. We especially like this tool for tight corners of the greenhouse. It tills best when pulled toward the operator. It is faster than the claw hoe. However, because it splays soil to the sides as it tills, it requires more raking after use. Third is a walk-behind tractor. We use a BCS 710 with 18-inch tiller attachment for blending one or two beds (BCS is the brand). Walk-behind tractors have two wheels, like the lower-cost Rototillers sold at hardware stores. However they are much more powerful because, like a tractor, they transfer power from engine to implement through a PTO (power take-off) gear shaft rather than a belt, which can wear down and slip.

As our beds are 30 inches across, two passes with the 18-inch tiller covers the entire width. We adjust the handlebars to an offset position while walking in the aisles to avoid stepping on the bed. To achieve maximum depth, for deeply blending in compost for soil building, we removed the depth-adjusting bar that came attached to the tiller. We found that the bar, even at its deepest setting, raised the tiller depth by a few inches.

Bigger is not necessarily better when it comes to walk-behind tractors. We chose this BCS 710 with 18-inch tiller because it squeezes nicely into small spaces, yet is powerful enough for tough soils. We use it to shallowly blend in composts and to deeply till if needed.

With tines lifted up, the BCS 710 also performs shallow tilling well. We tried shallow tilling with a tilther—a battery-powered shallow tiller without wheels—but sticks from our compost and rocks from our soil plugged it up. With the BCS 710 there is some compaction, but it is minimal and has never been a problem.

Our fourth blending tool, for larger areas, is the tractor with a 60-inch Woods rotovator. The rotovator is several times faster than our other aerating methods, and with its large tines it does a better job of homogenizing the soil. To minimize compaction from the weight of the tractor, we wait for relatively dry conditions. We also minimize compaction by keeping tires in the aisles. With this type of approach, compaction

Initial Soil Building with Compost

In 2016 we built up a plot with poor tilth, low organic matter, and slow drainage using black and brown composts in amounts based on testing and observation. These photos tell the story. For a brand-new plot, start by killing the sod, or whatever is growing on it, by smothering it for several weeks with tarps (see chapter 8), or by plowing. Chisel-plow and till if needed to loosen compacted ground. This work can be more demanding than you realize. While walk-behind tractor attachments, such as the rotary plow, can sometimes accomplish the job, tractors perform the work most efficiently.

To build up this plot, we hired a trucker to dump low-N leaf mold directly on the plot, which we then spread with a skid loader and by hand. Next we added high-N brown compost, chisel-plowed, and then tilled to break up the compost and blend it in to the depth of the rotovator, about 8 inches. A few months later we harvested abundant crops from the plot, which will remain fertile for several seasons.

has never been a problem for us. In fact, after eight years of using a tractor, our soils are looser than ever.

Another option for blending is to use a power harrow. In contrast to tiller tines, which rotate vertically, the harrow tines rotate horizontally. Power harrows leave soil layers more intact. They stir rather than flip. Also, with harrows weed seeds are more likely to remain buried rather than being pulled to the surface. Because there is no slicing of the subsurface soil, the harrow will not create a hardpan, as sometimes happens with a tiller. Power harrows are available for both tractors and walk-behind tractors.

Despite these advantages of the power harrow, we chose tillers over power harrows because at the time we were buying, the cost of a tiller was less than one-third that of a power harrow. We did not see the benefits paying off. Also, we actually want some soil inversion in order to blend compost, and to help break down and homogenize chunky composts and clay. If we are concerned about too much inversion, we slow down the tines by gearing up the tractor while lowering the rotations per minute (RPMs). However, both tillers and power harrows are found on successful farms, and if possible I would recommend trying each to see which works best for you and your soil.

Step 4. Shape

The final step is to shape the ground. We shape by creating raised beds: these allow excess water to run off and further increase soil depth. We only shape in the field, not in the greenhouse, where we can control watering, with the exception of greenhouse tomatoes—we raise soil for them slightly to give their roots more room. Shaping has been a critical step on our clay soil. If you are lucky enough to have beautifully drained fields, raised beds could still benefit your crops, but will be more optional.

In greenhouses, we create raised beds by hand for tomatoes. We have not found it necessary to raise beds for other crops in the greenhouse, as the ground stays soft and there is never excess rain. We do sometimes use a 12-inch flat shovel to scoop an inch or so of soil from the aisles to help define the beds for foot traffic. Also, simply by walking up and down the aisles we depress the pathways, leaving the beds slightly raised.

In the field we use a compact bed-shaper implement. While mastering this tool takes practice, the results are impressive. It will raise 6-inch-deep beds, moving thousands of pounds of soil in minutes while we sit comfortably in the seat of the tractor. Having shaped beds by hand for years, I was elated the first time I used this tool. Because the shaper works best on tilled ground, we usually use it after the rotovator.

Using a compact bed shaper to quickly create deep raised beds. The pan is slightly convex, so the bed is peaked in the middle for drainage. The shaper can also lay plastic and drip tape. We aim for 18 inches between beds. Note quick hitch for efficient and safe mounting.

This bed rake features 20 4-inch curved tines. The aluminum tines are not designed to grip soil; rather, they should ride almost horizontal to the surface, gliding rather than digging. Note the upright angle of the handle and "two thumbs up" hand position.

The No-Till Ideal

The ideal, of course, is no-till farming, eliminating the blending step and perhaps also aerating. This cuts out motion waste, saves fossil fuels, and could be healthier for the soil, although this is an area with much research still under way.

The definition of no-till varies widely. For some, no-till means never working soil at all except to put plants or seeds in the ground. For others it means only tilling or disking shallowly, rather than deeply. Tilling, in my mind, involves any activity

To clean beds and rotate from one crop to another we first mow crops if they are tall and then pull them out by hand or with rakes. In this photo I am slicing under spinach with a root digger to make pulling easier. We only use the tractor for this task in the middle rows of a greenhouse, where we can access them easily. We use wheel hoes for this task in harder-to-reach areas. In summer, when debris breaks down quickly, we don't usually pull crops but rather till and wait a few days or weeks to replant.

where rotating blades or plows touch the ground, no matter how deep they go.

In conventional agriculture, no-till methods involve killing old crops and cover crops with glyphosate, the key ingredient in the weed killer Round-Up, and then drilling seeds into the stubble. Some organic methods include killing cover crops with a crimping implement and also removing old crops by hand—or killing them with plastic sheets—rather than incorporating them with a tiller.

My view is that while no-till represents an ideal, at least some aerating and tilling is still practical and essential on the market farm. At a minimum, almost all soils need an occasional aeration through broadforking or some other means to loosen compaction from rains, and the top 1 or 2 inches must be fluffed through light tilling for direct-seeders to work properly. Tilling breaks up compost chunks and prepares ground for hand transplanting and bed shaping. It also represents an efficient way to turn in old crops. In addition, tilling prepares soil for the paper pot transplanter (see chapter 6).

To still harness the benefits of no-till, we do apply these approaches:

1. *The rule of minimums.* To save moves, we take a suspicious approach to each of the four steps. Do we really need to amend, aerate, blend, or shape every time we plant?

   The answer is no. We often skip steps. For example, in many cases, after pulling an old crop we only aerate and skim-till the top inch or two of soil to prepare a bed for planting. Sometimes we skip all four steps—we just pull and plant. Analyzing the necessity of every move, rather than sticking to a rigid formula, cuts enormous motion waste from our farm. A few cases in point from 2017:

- In mid-May we harvested our first crop of baby head lettuce from a bed in one of our greenhouses. At the same time, I had a tray of romaine heads in paper pot cells, ready for transplant. I skipped all four steps and simply transplanted them into the same bed, offsetting the rows a bit to avoid old roots, which we left in the ground.

- In early September we pulled two beds of basil from a greenhouse, replacing them with direct-seeded baby greens. Baby greens are light feeders, so we did no amending, and because they are relatively shallow-rooted we also skipped the aerating. All that was required was to clean the area, shallowly till the top 2 inches, and seed.

2. *Multiyear plastic.* Because compost provides long-term nutrition and soil structure, we can often pull crops out of plastic mulches and simply refill the holes with new plants. This saves the work of shaping beds and laying new plastic and drip tape for each crop. It is the leanest way I know of to grow many transplanted crops such as kale, chard, leeks, bulb fennel, kohlrabi, and onions. In some cases, we "pull and plant" six crops in succession before replacing mulch. I credit Pete Johnson at Pete's Greens in Craftsbury, Vermont, for this

We grow some crops with virtually no bed preparation at all using multiyear plastic. In this photo, Swiss chard, leeks, and onions were planted in holes vacated by lettuce and kohlrabi. We sometimes add a pinch of compost into the holes as we transplant. We also use this "pull and plant" approach on beds without plastic.

idea. In greenhouses on his farm, they transplant winter crops like bok choy and napa cabbage into plastic left in place from summer melons. The best material for this system is thick plastic, which resists tearing over multiple seasons, or landscape fabric, which also can be used for many years. We only use plastics on crops that are in the ground for two months or more (see chapter 5).

The key to success—once again—is observation. At the time of removal does the preceding crop show excessive yellowing in the leaves? This might indicate nutrient deficiency and a need for amendments. On the other hand, if you plan to grow a light-feeding crop such as head lettuce and the preceding crop was robust, with dark green growth, why bother amending, aerating, blending, and shaping? Just pull and replant.

Over time, beds flatten out with wind and rain. We don't bother to reshape beds every time we plant because we don't mind if the beds are a little flatter in summer when excessive rain is rare. With proper equipment—tractor and bed shaper—reshaping the entire farm takes just a few hours.

If we are transplanting, we simply push plugs into the shaped beds—or run a paper pot transplanter—immediately after bed shaping.

If needed, we use a 30-inch bed preparation rake to further level the ground and remove trash. We always rake before direct-seeding in order to minimize problems with the seeder. Our bed rake is specially designed for this purpose (for information on suppliers, see appendix 1). Standard landscaping rakes, while stronger, do not pick up debris as effectively.

Soil preparation is perhaps the most personal of farming tasks. Touching soil is a visceral act, and preparing soil literally grounds farmers, connecting them to nature. We strive to improve our soil efficiently by analyzing our movements so that we can quickly get back to seeding, harvesting, washing food, and other tasks in the value-adding category.

We often walk our fields slowly to simply enjoy the beauty of our farm and of nature. It is important not to skip these occasional or scheduled walks. Many people envy farmers for their idyllic work environments. Sometimes it behooves us to slow down and appreciate ours.

CHAPTER 3

Compost Making, Small Farm–Style

There are four purposes of improvement: easier, better, faster, and cheaper. These four goals appear in the order of priority.

—Shigeo Shingo

The only known record of vegetable-growing practices used in medieval Japan is found in a document called "Regulations of the Engi Era," issued in 927. The regulations dictate in detail the materials, steps, and time needed to supply the reigning court with vegetables. For example, 1 *tan* (.3 acre) of spring onions required the following:

4 *shō* [3.4 liters] of seed
1,200 seedlings
1½ days for three tillings with 1 ploughman
1 ox driver and 1 ox
1 day for hand tillage
2 days to form ridges
35 days to transport 210 loads of manure
½ day for sowing (in the 8th month)
20 days for transplanting (in the 2nd month)
10, 9, and 7 days for three weedings

Notice the time devoted to moving manure: of 87½ days' work, 35 days—almost half—are devoted to the task! This wasn't just the case for spring onions. Here is the protocol for 1 *tan* of coriander:

2 to 5 *shō* [21.3 liters] of seed
1 day for 2 tillings with 1 ploughman
1 ox driver and 1 ox

2 days for hand tillage
2 days to form the ridges
22 days to transport 132 loads of manure
½ day for sowing (in the 3rd or 8th month)

For coriander, of 28 days' work, 22 are spent moving manure![1] In fact, for almost every crop listed in the "Regulations," the bulk of effort required was to move manure and compost.

F. H. King, in his journey in the early 1900s to discover how farmers in China, Korea, and Japan achieved high yields for centuries without depleting their land, also found a staggering commitment to using manure and compost. He saw pits for preparing composts and "piles of stable manure awaiting application, all of which had been brought up the slopes in baskets on bamboo poles, carried on the shoulders of men and women."[2] He observed a field of leeks onto which workers had applied 16,000 gallons of manure per acre.[3] He calculated that in 1903 the Japanese cut 2,552,741 acres of green manure from un-farmable "hill lands," yielding 7,980 pounds per planted acre, replacing the need for 46,800 tons of rock phosphate.[4] In 1908 King wrote, "Japanese farmers prepared and applied to their fields 22,812,787 tons of compost . . . from cattle, horses, swine and poultry, combined with herbage, straw and other similar wastes." He concluded, farmers "have long realized that much time is required to transform organic matter into forms available for plant food. By such practice, with heavy fertilization . . . the soil is made to do full duty throughout the growing season."[5]

◼ ◼ ◼

When we first started to rely heavily on compost for our soil fertility, we were unique in our area. Many farmers use compost, but not in the amounts we were using.

Why has this ancient practice diminished so markedly? For one, since the advent of the railway system, farmers have had other options. When railroad transport came along in the 1800s, companies could ship mined minerals, and eventually chemical-based fertilizers, across thousands of miles. A more recent reason is that farming has gotten too big, and labor too expensive, to follow older practices. On hundreds of acres of land, it is impractical, even with large machinery, to imagine a system relying on the regular application of compost. It is much cheaper to distribute lighter-weight and less bulky granular and liquid fertilizers. In terms of transportation costs, composting is better suited to small farms.

Even so, what small farmer has 22 out of 28 workdays to devote to making and moving compost? In our earliest experiments we formed piles by hand, layering straw, manure, and garden scraps. We framed the piles with straw bales. We hauled everything with pitchforks and garden carts. Soon we realized that our backs would not hold up, and that we couldn't afford the time to make the amount we needed. If we wanted to get serious about using compost, we had to lean up.

If you are fortunate enough to live near a composting service that will give you a good price for a quality product, by all means save your time and buy from them. In many places, however, such a product is not available. We rely both on purchased compost and our own, which we make on the farm.

Shigeo Shingo, one of the designers of the lean system, writes that

Compost keeps soils teeming with life.

improvement involves four purposes: "easier, better, faster, and cheaper." Our compost making aims to achieve all four. To summarize our system: We hire trucks to deliver raw material (old hay bales, straw, leaves, manures—we are not too picky) from wherever we can for as low a cost as possible in the spring, aiming for around 75 cubic yards. Then we form a windrow about 6 feet tall by 9 feet wide by 50 feet long with our skid loader, a much faster method than hand-forming piles. Throughout the season we turn it five times (the easiest way is to move the whole pile) and monitor to ensure the compost reaches a temperature of 131°F for 15 days. This high temperature produces the best-quality compost because it kills weed seeds. We water it in the summer so it doesn't dry out. We cover it in the winter with a compost fabric so it doesn't become waterlogged from rain. Under ideal conditions, the pile is ready to use in four to five months, but we typically save it for the following season, which begins in January in the greenhouse. With practice, compost making does not consume a lot of our time. Here are the steps to make compost in detail:

1. Gather raw materials.
2. Build the pile.
3. Turn the pile.
4. Regulate moisture and temperature.

Step 1. Gather Raw Materials

Among the most important tasks in the first years of your farm, if you plan make your own compost, is to find sources of raw material. Unless you have haying equipment and dozens of extra acres, or a feedlot of animals depositing manure, this means searching off-farm.

You might be surprised at what is available. Every year our compost changes because our raw materials come from so many different places. We have used the following:

- Moldy hay from neighboring Amish farmers
- "Floor bales," bales of hay stored on the bottom layer in a barn, often too poor in quality for animal feed
- Large, round bales of straw that have sat out in the rain
- Cuttings from our pasture, collected with a flail mower and catch basin
- Leaves from our farm or trucked in from other properties
- Grass clippings from our farm or trucked in
- Vegetable wastes from our farm and other households
- Duck manure (from an organic farm) in sawdust bedding
- Duck manure in straw bedding
- Chicken manure from our own chickens

Two Essentials for Making Compost

To make sufficient compost for a small farm you will need a low-cost supply of raw materials and a skid loader or tractor with a bucket. In our experience it is not possible to make enough compost for a commercial farm by hand. If you can't find or afford these essentials, I still recommend making a smaller amount of compost by hand, turning your field scraps into a product you can use, using steps 1 to 4 outlined in this chapter. You will generate the scraps anyway. You might as well put them to use, even if the amount of compost you make is not sufficient for all of your needs.

We cast a wide net because the best composts are made of a broad mix of materials. As long as a material was once alive or from a living thing (such as manure) it will decompose, and we use it, provided it is safe. With the

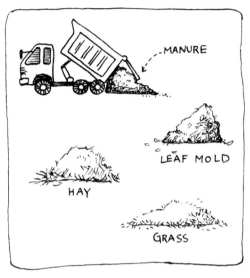

1. Gather local raw materials in the spring. A mix is best.

2. Build a loose pile. A good size is 9' wide × 6' tall × 50' long.

3. Turn the pile at least 5 times to homogenize, add oxygen, and even out moisture.

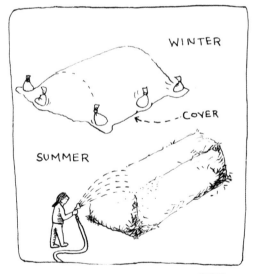

4. Regulate temperature (15 days at 131°F) and moisture. Water to retain moisture, or cover to prevent saturation.

Four Steps to Compost Small Farm-Style

Compost Recipes

Simple Low-N Compost
FOR C
2 scoops old hay (mostly dry grass) or straw
2 scoops leaves

FOR N
½ scoop manure, food waste, or grass clippings

Simple High-N Compost
FOR C
1 scoop straw (can be bedding
 in the manure)

FOR N
1 scoop poultry manure

Optional: Add ¼ scoop soil or finished compost to help activate the heap.

Contamination in Compost Materials

In some ways, compost piles act like a wetlands, filtering pollutants from the environment. Still, some pollutants should stay out of a compost heap.

In particular, a class of herbicides from the picolinic family can persist on vegetation and in the soil for months and years. They do not break down in the compost process, and they kill crops. These persistent herbicides are used on golf courses, hayfields, grain crops, and lawns, among other places. There are four types: Clopyralid, Aminopyralid, Aminocyclopyrachor, and Picloram.[6] Testing for them is expensive and difficult, though it can be done. Consult your local extension agent, who can direct you to testing labs. It is best to trust your source as to the presence of contaminants. Contaminants can also show up on municipal yard wastes (in general leaves are safer than grass, since they are unlikely to have been sprayed) and in animal manures (through digest-ed food or administered drugs), among many other places.

Those who make compost draw lines at different places when it comes to contaminants. At one extreme, some require all-organic sources of raw materials or use only material from their own property. Others use anything that comes their way, on the theory that nature breaks down GMOs and nearly all contaminants (except for those persistent herbicides) in the composting process. We take a middle-of-the-road approach. We accept non–organically certified leaves and pasture cuttings, as long we know they were not sprayed with persistent herbicides. But we usually choose organic sources of manure. We also steer clear of municipal sewage wastes, called *biosolids*, though these are now commonly used on larger farms in our area, because of possible contamination from treatment chemicals. If your farm is certified organic, you may want to consult with your certifier about your raw materials.

widespread use of antibiotics, herbicides, and GMO feeds, what constitutes "safe," of course, varies from farmer to farmer. It is up to you to decide your comfort level with raw materials. (See also the "Contamination in Compost Materials" sidebar on page 48.)

As explained in the previous chapter, we compost animal manures and their bedding separate from green manures. An ideal mix contains 25 to 30 parts carbon (C) to 1 part nitrogen (N), although we do not take the time to calculate C:N ratios—quality composts can result from a wide range of ratios. The key is to avoid extremes. Piles too high in N will become too hot, killing the beneficial microorganisms. Piles with too much C and not enough N will not heat up enough. As a general rule, manures and green grasses are "hot" (high-N) and materials such as straw, leaves, and wood chips are "cold" (high-C). For organic certification, you may need to document C:N ratios. One good online source of carbon-nitrogen information is the Center for Environmental Farming System's "Composting on Organic Farms" (see "Composting Resources" on page 222).

Our biggest challenge is price. In many areas, composting ingredients can be had for free. I know of one farmer who is paid to pick up manure. Unfortunately, that's not the case for us. We secure most manure for free, although we pay for delivery. Our cut-off point for what we are willing to pay for raw material, without delivery, is around $6 per yard. We are not concerned about weeds—our piles will heat up enough to kill them. Wood shavings are fine, but

Applying Composted versus Raw Manure

If you have extra land, one option is to skip composting and apply manure directly to your fields, as long as you apply it well ahead of when you plan to grow a crop and blend it in. The manure will stabilize on its own and rid itself of pathogens. For organic certification, the time required between application and expected harvest is 120 days for crops with leaves or fruits that touch the ground, or 90 days for all other crops.

If you apply fresh manures directly to crops as a side dressing, you risk "burn-ing" plants through excessive nitrogen, or through ammonia gases released from the manure. Applied to the soil too close to seeding or transplant time, fresh manures can kill seeds and transplants, and pathogenic microorganisms can find their way onto food.

On our farm, we choose not to use raw manures because our plots are almost always in production. By composting the manures first, we enjoy the flexibility of applying them when we need them.

Buying Partially Finished Compost at a Lower Cost

When applying compost to the field, you do not need a perfect blend. Slightly unfinished compost actually delivers nutrients over a longer period of time because they have not yet converted to stable forms. Only when used with potting mix do you need fully finished, sifted compost.

When we explained our needs to a composting facility in our area we were able to secure a much lower price for rough, unsifted compost. The mix often arrives with leaves only halfway decomposed and small sticks. It sometimes clumps together. It is not a mix into which we would directly seed. But blended with our soil it contributes texture, a large amount of organic matter, and a wide range of nutrients—released slowly over time—for our crops. After a

When purchasing compost you do not need a refined mix unless you are using it in potting mix. We negotiated with a supplier to lower cost.

season or two, the sticks and larger pieces disappear and leave behind a mellow, loose soil.

we don't allow large wood chips. Other than that we are not concerned about particle size. Even clumped-together hay soon breaks down and separates.

Step 2. Build the Pile

When trucks arrive, I direct them to our compost yard, set away from our greenhouses and outdoor growing area to prevent leaching. The yard is easy for dump trucks to back into. It is surrounded by grass to keep runoff to a minimum. It is out in the open, as there is no need for a building.

To build the piles, we push the raw materials together with our skid loader to form windrows 9 feet wide by 6 feet tall. We mix the different materials together as we build the pile. We used to place bales of straw around the perimeter for containment and insulation, but we no longer do this: it adds cost and makes the pile inconvenient to turn.

There is no consensus on the ideal size of a compost pile, but in general bigger is better, as long as air can reach the core. Most

After initially building soils, a small farm does not need much compost. Our compost piles are 9 feet wide by 6 feet tall, and 50 feet long. One or two piles per year suffice for our needs. We build piles one year in advance of using them.

decomposing activity happens in the core—in the middle—where heat builds up. You want a large core. Also, small piles can dry out quickly because they have a high surface-to-air ratio and do not insulate themselves as well as larger piles. I credit Steven Wisbaum, a professional compost specialist at CV Compost in Charlotte, Vermont, for helping us design our system. Wisbaum notes that, from his experience, naturally porous piles made with low-density materials such as horse manures or shredded debris can be built up to 15 feet wide. Piles larger than that are impractical to manage and can suffer from a poor rate of gas exchange, trapping gases that need to escape from the core. When using denser materials, such as leaves or grass clippings, I recommend a smaller pile so that it "breathes" better.

In actual practice, for most farms the size of the pile is determined by equipment. Ours is determined by our Gehl 44-horsepower skid loader, which can comfortably dump loads at a height of 6 feet. If you use a compost turner, a tractor-powered implement for mixing compost, your piles might be shallower and narrower.

For length, I recommend piles 50 or 100 feet long, as most compost covers are sold in 50-foot increments (discussed further in "Step 4. Regulate Moisture and Temperature" on page 154; see also appendix 1 for more information on compost covers). Piles should also be sited near a source of water for irrigating. If composting raw manures, consider your neighbors and minimize foul odors by mixing fresh manures with other materials. Alternatively, cover or mist fresh manures, or use an additive that discourages the growth of odor-producing microbes.

Step 3. Turn the Pile

There are three reasons to turn piles:

1. *Homogenize* the compost, blending raw ingredients together and breaking up clumps. This makes more food available to microbes as decomposition proceeds.
2. *Restore pile porosity* lost through settling, thus allowing oxygen into the core. While some oxygen is introduced during the actual turn, most oxygen used in composting arrives through "passive aeration," as air moves in and out through pores.
3. *Even out moisture* in the pile, mixing drier outer layers with the wetter core.

The net result of turning is heat, a primary by-product of decomposition. Sustained heat above 131°F kills off pathogens and weed seeds and also stimulates faster decomposition. We turn a pile by picking it up bucket by bucket and shifting it over by about 10 feet, in effect building a new pile.

While we rely on a skid loader, professional composting companies and large farms use compost turners, which require large tractors. This machinery costs tens of thousands of dollars and does not pay off for most small farms. Larger operations also use grinders and sifters that break apart or separate larger debris. We have never used either one and have not had problems, since we do not need perfect-looking compost. If larger clumps survive until planting time, we can pull them out with the 30-inch bed rake or break them up with a tiller.

Another common practice is to use a manure spreader to turn piles. With this system, compost is scooped up with a skid loader, dumped in the spreader, and emptied into a windrow. The process adds lots of oxygen and breaks the compost into small pieces. While we have done this in the past, we don't find the spreader to be necessary for our current mix of ingredients.

Composting as Tiny-Animal Husbandry

When making compost, I think of myself more as an animal farmer than a plant farmer. That is because the basic task of composting is to create ideal conditions for tiny organisms—aerobic bacteria, fungi, actinomycetes, and other micro-scopic beings that turn raw waste into priceless humus. Raw materials give them food. Turning gives them fresh air. Irrigating gives them water. Then I step out of the way and let the tiny animals get to work.

Is there an ideal number of times to turn a pile? Many educators advocate for frequent turnings, as often as once per week, on the theory that more turns equal more oxygen and heat. Others argue for a "minimal number of well-timed and thorough turns." Steven Wisbaum of CV Compost says this saves labor and equipment costs, reduces the release of foul odors, conserves nitrogen and moisture, and produces compost with higher organic matter content.[7] With this system, a farmer uses proper management—building porous piles with a wide variety of materials, and

Composting provides plenty of fertility for close-cropping systems.

maintaining even moisture—so that microbes, not machines, do most of the work. As we are certified organic, we stick to the five-turn approach required for certification, while attempting to use good management practices to keep turns limited to five. Once a pile is complete, there is no need to keep turning it unless it was allowed to dry out, in which case turning, plus irrigation, can bring it back to life.

An alternative to the turned-windrow method is to create static aerated piles. This involves inducing oxygen into piles through tubes that force in air. Several websites offer plans for these systems, which can be a good option if you don't have turning equipment, though these systems will involve a more complex setup and more maintenance.

Step 4. Regulate Moisture and Temperature

Just as the tiny organisms in the compost heap need food and fresh air, they also need water and comfortable temperatures.

1. *Moisture.* While regulating water content is essential to compost making, many farmers overlook it. To monitor moisture, we pick up a handful of compost and examine it. With a hard squeeze the mix should just barely stick together, and water should drip out, indicating moisture content of around 60 to 75 percent. If water drips out with no squeeze or a light squeeze, it is too wet. If we see no droplets, it is likely too dry. Several instruments are available for more precise moisture reading, but in our experience they are not necessary.

   If there is too little water, the composting process stops. In the summer we water compost with a spare 2-inch drip tape main line run across the top of the pile into which we have inserted micro sprinklers every 3 feet. If we have several piles to water, we use overhead sprinklers to water them at the same time. It is important to water as evenly as possible across the entire surface. Drip tapes do not work well for watering evenly because water tends to pool at the emitters.

   If compost is too wet, decomposition also grinds to a halt because pore spaces, where oxygen enters and leaves, become plugged up. In our location, heavy late-fall rains and winter snowmelt can saturate our compost. For that reason we cover the pile with a composting fabric from November to March. The cover is made out of an advanced-technology fabric. It is porous so some rain and snowmelt can pass through and CO_2 can still escape, but once saturated the pores close, maintaining even moisture. The cover also helps insulate the pile. (See "Propagating and Trans-planting" in appendix 1 for suppliers of compost covers.)

2. *Temperature.* Proper temperature is as important as proper mois-ture level. We record the temperatures of our piles on a spreadsheet before each turn. Long-stemmed compost ther-mometers are the easiest way to measure the temperature.

If the pile is not heating up, we turn it more or add more nitrogen, usually in the form of fresh duck manure. We might also adjust the pile's moisture content, as soppy mixes and bone-dry mixes do not heat up well. The ideal heat, as men-

Tubing with emitters to water compost. Watering compost is essential but often overlooked.

tioned, is 131°F. Piles that are too hot, above 155°F, suppress certain beneficial microbes and need to be cooled down. We just leave the pile undisturbed, allowing it to cool naturally.

We know our compost is finished when it turns dark brown or black, smells earthy, and crumbles easily (resembling cake crumbs) and when raw materials are no longer distinguishable from one another. Even after it is finished, we try to maintain even moisture to keep microbes alive, though we allow it cool below 131°F.

In most cases, we then pick the compost up with the skid loader and move it to where we want to use it. In the case of tomatoes, as stated, we sometimes mix in minerals with the compost for our monthly side dressing, depending on the result of tissue samples and soil tests. To do this, we spread out an 8- to 10-inch layer of compost on the ground somewhere, sprinkle on the mineral powder, and till with the tractor or BCS. In addition to blending in minerals, this also makes for a fluffy mix that is easy to apply by hand around the base of the tomato plants.

Spring on the farm. Early in life plants require tender care to prevent defect waste. Our goal is for every seed to germinate.

CHAPTER 4

Successful Seed Starting

Ask "why" five times about every matter.

—*Taiichi Ohno*

In 1924, Sakichi Toyoda, an engineer from Kosai, Japan, invented a new type of spinning loom. The machine, called the Toyoda Automatic Loom Type G, featured a seemingly simple innovation: when it found a mistake, it stopped working. On other looms, yards of materials might be wasted before a human operator spotted an error.

A decade later Toyoda Loom Works spun off the automaker Toyota. Toyota used the same approach: engineers designed systems to stop defects fast, before mistakes could compound and cause waste. At first, to halt production, workers who spotted a mistake pulled cords called *andon cords*, conveniently placed above workstations. Taiichi Ohno, who helped implement the system at Toyota, encouraged cord pulling: "Workers should not be afraid to stop the line," he said.[1] Now lasers, in addition to workers, often detect errors.

Defect waste is perhaps the most ubiquitous type of waste in vegetable growing. There is no way to completely eliminate defect on a farm; nature resists that. However, you can ask: when in my timeline, from seed to harvest, does defect tend to occur most frequently? And then focus there.

When we started to apply lean, the answer was in the first few weeks of our plants' lives. That is when plants are the most fragile, requiring delicate care, and when much can go wrong. To cut out defect waste, we set a goal: get every seed to germinate. We set about improving our seed starting, asking "why?" every time a seed didn't germinate. According to Taiichi Ohno, it takes asking "why" five times to get at the root of most problems. We don't get every last seed to germinate, but with lean thinking we now come close.

Germinating in Chambers

Vegetable seeds need two conditions in order to germinate:

1. The right temperature
2. High humidity

In theory, one could germinate a seed in thin air, as long as that air was the right temperature and very humid. Soil is just a landing pad for the first root (called a *radical*) that emerges from the seed. Light isn't usually needed until the first tender leaf (called a *cotyledon*) starts reaching for the sun. The key to consistent germination, we have learned, is to supply temperature and humidity precisely.

Our favorite germinating method is to use germination chambers, insulated high-humidity boxes heated with a pan of water. The chambers supply both heat and humidity with remarkable consistency. With this high humidity method, crops also germinate faster than in less-controlled environments. The chambers are space-efficient: 30 flats germinate in the space normally required by one because the trays are stacked. I referred to our use of the chambers in *The Lean Farm*.

We keep two chambers in the front part of our propagation greenhouse. One is an old bakery cart onto which we have glued rigid insulation. We keep it set at 68°F for cool crops such as lettuces. The other chamber is a chest freezer set upright. We keep temperatures there at 78°F or 85°F depending on the crop (see the "Getting Temperature Right" sidebar on page 59).

An alternative is to dedicate a closet or a small room to germination. A larger-scale grower near us built an insulated 8-by-8-foot room attached to his greenhouse. To heat it he used a propane wall-mounted blue flame heater. His seedling trays sit on easy-to-reach shelves.

Our germination chambers hold more than enough flats for our needs. We stack trays one on top of the other, with no spacers in between, to economize on space. We can stack about 10 flats on top of one another before they start to feel wobbly. As long as we remove trays on time, this has not caused problems. We use leftover greenhouse parts as shelf supports.

Because there is no light in the chambers, crops must be removed as soon as they germinate and be placed onto greenhouse tables. Wait too long and they will turn into a tray of noodles. We check the chambers every morning and evening. Some growers install a window in the front door for easier viewing. We prefer an insulated door to keep temperatures even. Not all seeds germinate at the same time, of course. We pull a flat from a chamber when we see that the first few seeds have popped. The rest will quickly follow suit on the tables.

Getting Temperature Right

All vegetable seeds have an optimum temperature for maximum germination. For beans, it is 80°F. For squash, 95°F. For lettuce, 75°F. For the sake of simplicity, we divide our crops into three heat categories—68, 78, and 85.

Table 4.1. Temperature Settings for Optimum Germination

Heat Category	Vegetable Seed
68°F	Lettuce
	Peas
	Spinach
78°F	Beans
	Brassicas (kale, broccoli, kohlrabi, etc.)
	Carrots
	Swiss chard
	Onions
	Radishes
	Tomatoes
85°F	Beets
	Cucumbers
	Eggplant
	Peppers
	Squash, summer and winter
	Turnips
	Watermelon

Whether growing plants on a heat mat or in a germinating chamber, we aim for these temperatures. If direct-seeding, we often wait for soils to reach these temperatures before seeding to increase the odds of germination. For example, we do not direct-seed spinach (in the 68°F category) into soil that is warm to the touch, or that measures higher than 75°F with a soil thermometer. Instead, we germinate plants indoors in multicell plugs for transplanting later.

In general, it is riskier to supply too much heat than too little. For example, tomato seeds will often germinate in 65°F soils, albeit at a lower rate than optimal. But above 95°F they will cook and none will germinate. For us, lettuce seed direct-seeded into very cold ground, such as in a winter greenhouse, will eventually germinate in a few weeks. But in very hot ground, above 90°F, few if any seeds will sprout. For this reason, our germination chambers are never hotter than optimum. In July and August, when greenhouse temperatures rise above optimum, we germinate seeds in our basement using a grow light (see "A Simple Grow Light Setup" on page 66). For a comprehensive list of optimum temperatures for vegetable seed germination, I recommend the University of Oregon Extension's "Soil Temperature Conditions for Vegetable Seed Germination" (link in "Small-Farming Practices" on page 221).

Plans for a Farm-Built Germination Chamber

Parts Required

Insulated airtight box, such as an old, nonworking upright refrigerator or a chest freezer set on end. These can usually be picked up for free or at low cost. The size is up to you, depending on how many flats you plan to germinate at a time. Allow for a few inches around trays for air to circulate. The box should not be wooden, to avoid mold.

Rustproof pan for heating water, 15 by 20 by 8 inches: **$50**

Metal electrical box: **$2**

Hot-water-heater element, 1,440 watts, 120 volts: **$16**

This germination chamber is a repurposed upright refrigerator. A thermostat is mounted on the outside, with a probe hanging inside. A pan of water, heated by an electric heating element, sends steam into the chamber, creating ideal conditions for germination.

Flange for hot-water-heater element: **$5**

10-foot length of #14 wire rated for exposure, such as SEOW, or standard Romex wire installed in electrical conduit: **$10**

Remote sensor temperature controller, such as a DuroStat Watertight Thermostat with Remote Sensor: **$50** (To keep temperatures from fluctuating, it is critical that the temperature controller feature the ability to set a differential of fewer than 3°F.)

Metal switch box and 120-volt switch: **$5**

Three-prong #14 cord, to run from switch box to 120-volt outlet: **$10**

Optional: stock tank float valve: **$20**

If you find a free throwaway refrigerator or chest freezer, the total cost for parts comes to approximately **$168**.

Assembly Instructions

1. Place empty box on a level surface. Make sure door opens freely. An ideal location is in your propagation house, as this allows for quick transfer of trays to growing benches. In summer, when greenhouses can overheat, consider placing the chamber in a cooler conditioned space, such as your basement or a processing room.

2. Drill a ¾-inch hole in a corner of the floor of the box for drainage.

3. Place rustproof pan in the middle of the floor, being careful not to cover the drainage hole. The pan should have an electrical box welded to the outside with a hole sized to your hot-water-heater element, plus a hole with a plug in the bottom for drainage. You will want to replace the water in the event

a lot of dirt falls into it. An option to prevent dirty water is to place a screen over the pan, though we have not found this to be necessary.

Unless you are handy with a welder, I recommend hiring a machinist to fabricate the pan. We have used salvaged stainless steel baking dishes as pans, as well as custom-made pans with a galvanized coating.

4. Mount remote bulb temperature controller to outside of box.

5. Drill ¼-inch hole and run temperature-sensing bulb into the chamber, near the middle. This will signal the element to turn on or off. Set the thermometer to the lowest possible differential setting; this will keep the temperature range small.

6. Mount switch box and switch.

7. Wire up the chamber:

    a. Run #14 wire from element to temperature controller.

    b. Run #14 wire from temperature controller to switch box.

    c. Wire cord to switch box.

I recommend consulting an electrical contractor for this portion to make sure the system is sized properly and well grounded.

8. Optional: Mount stock tank float valve to rustproof pan. Connect hose with low-pressure water to the float valve. The reason for a float valve is to guarantee a supply of water. A hot-water-heater element exposed to air will fry and need to be replaced. To reduce parts, we no longer use a float valve. Instead, we fill our pan by hand about once per week.

Potting Soil and Plug Flats

We find that it is not worth the effort to make our own potting mix. Purchased mixes are tested, reliable, and readily available. For us, they have been worth the added cost. Compost-based mixes have provided the best results for our plants. We prefer a lightweight mix for germinating and a denser mix for potting (for companies offering compost-based mixes that I have used, see "Propagating and Transplanting" in appendix 1).

We generally buy by the yard, in sling bags, and make sure to stock up for the next year by late fall. We store our potting mix in the greenhouse where it won't freeze, and we keep it evenly moist but not soggy. It should fall apart when you lift it but still feel wet.

We start seedlings in commercial-duty, reusable plastic trays, called *plug flats*. The flats hold water well and are easy to move around. We use four sizes (see table 5.2 on page 75). We fill the flats by hand, making sure each cell is full but not overly packed. To speed the job we use a 12-inch square-point shovel to dump potting mix into the trays and then brush excess off with a 6-by-14-inch scrap of rigid polycarbonate. To seed the flats, we make a ½-inch divot in each cell, drop seeds in by hand, and then cover

Storing Seeds

Even if you get heat and humidity right seeds won't germinate unless they are viable. The older the seed, the less viable it becomes. Even old seeds that germinate sometimes show a lack of plant vigor. Also, seeds stored in conditions that are too hot or humid will lose viability.

We keep our seeds in airtight plastic containers stored in our cool basement. We use eight containers, each holding a class of seeds—salad mix seeds, alliums, nightshades (tomatoes, peppers, eggplants), and so on. In the containers we keep silica gel packs, saved over the years from packaging, as desiccants.

Even so, many of our seeds lose viability after a season, and we rarely keep them for more than one year. To keep our seed inventories low, we purchase seeds in small batches as we need them throughout the season instead of putting in one large seed order at the beginning of the year. Vegetable seed companies invest enormous effort in maintaining vigorous seed, spending thousands of dollars on seed testing and on storerooms with expensive climate control equipment. We are happy to leave those costs to them.

A farm-made dibbler for punching ½-inch divots in a seedling flat.

with fine vermiculite, which allows air to reach the seed while keeping it wet. The vermiculite is also easy for seedlings to push through. As a final step, we lightly water with a fine-mist watering wand. Our favorite is the Wonder Waterer wand.

The Seedling Greenhouse

After germinating, seeds should go into a well-lit and heated area until they are ready to be transplanted. This area is often called a propagation greenhouse, transplant greenhouse, or seedling greenhouse. On a very small scale, you can set flats in front of south-facing windows in your house, perhaps supplemented with grow lights. We keep our propagation area heated to 60°F. For ventilation, we use automated louvers that open at 75°F, and endwall fans that kick on at 80°F.

Since this space is heated, it should be just the right size. We use a portion of a larger greenhouse as our propagation area. From January to March we divide it with a large sheet of plastic. The front section is heated and is devoted to our early starts. We can move the plastic forward or backward to adjust the size of the heated area. Eventually, as the weather warms, the sheet comes down and we heat the entire greenhouse.

The propagation greenhouse should have two layers of plastic, with air blown between the layers. Your greenhouse supplier can help you with this setup, and also with choosing the right type of plastic. Greenhouse plastic is a fast-developing field, and new plastics can save on heating costs and reduce condensation.

For heat I recommend starting with standard propane- or natural-gas-powered unit heaters that blow hot air into the space. These heaters represent a cost-effective, proven technology. They usually hang from greenhouse rafters and are ventilated through an endwall. Some growers attach perforated tubing to their heaters. The tubes run the length of the greenhouse, either in the air or under growing benches, and help distribute heat. We have not found tubes to be necessary. Consult with your greenhouse supplier to choose the right-size heater.

Growers are a creative lot, and many have developed more experimental heat sources, such as capturing heat from compost piles (tubes running through the piles that transfer heat to a radiator inside the

A greenhouse-within-the-greenhouse makes it easy to heat a small space in midwinter. This mini-greenhouse consists of a simple wooden frame built on top of two greenhouse benches. Large windows open for access. The roof lifts for ample ventilation. A 30,000 BTU blue flame heater, placed under the benches, warms the space.

greenhouse), or borrowing heat from in-ground tubes. Another approach is to heat growing benches with hot water tubing. This puts heat under flats where it is needed. These are more sophisticated setups to try after you have more experience.

Your seedling greenhouse should also have a horizontal airflow fan to circulate air, which helps drive moisture off plants, and the ability to ventilate with roll-up curtains or automated louvers. In general, the more fresh air, the better.

In our seedling greenhouse we also use a small plug-in hot water heater to warm water and avoid shocking tender starts as we water them. This is especially important when watering tomatoes and peppers.

I recommend planning for power outages or heater failure. We keep a small generator handy in case an electrical outage occurs. We installed a whip onto our unit heater that allows us to plug it into the generator. We

Our waist-high benches are supported by cinder blocks, which makes them modular. We use the front portion of a 22-by-70-foot greenhouse as a germinating area. We divide the greenhouse with a sheet of plastic if we don't want to heat the entire space.

Using *Heijunka* (Load Leveling) to Pace Your Seedling Work

At seeding time in the spring it can be easy to pile on work. It might seem urgent to fill your propagation house all at once with hundreds of trays of vegetables. But try to spread out the workload. On our farm we aim to seed about 10 trays per week, on Monday afternoons, from January to April, and a smaller number the rest of the year. The task takes a few hours. Because we repeat the same steps every week, it is easy to enter into a flow. We look forward to seeding time because we find it relaxing. As a guide, we use our *Heijunka* Calendar and *Kanban* Map, which we post on a germination chamber.

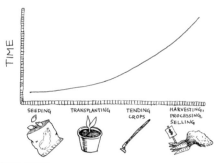

If you don't have time to seed it, you won't have time to tend it later. Baby plants require relatively little time to manage; older plants must be trellised, managed for pests, cultivated, and harvested. *Heijunka* at seeding time is the best way to prevent burnout later in the year.

also installed a smaller backup heater—a wall-mounted blue flame heater—in the greenhouse.

Finally, to support trays, you will need to construct waist-high benches with a perforated bench top. We use heavy-gauge steel mesh sheets stapled to a 2-by-4-inch frame. Whatever material is used, it should be sturdy—sagging benches will make uniform watering difficult. For ease of reaching, tables should be 6 feet deep or less. Our tables are 6 by 6 feet. We use cinder blocks to support the frames, allowing us to take the setup apart once each year for cleaning. We keep tables away from the edge of the greenhouse, where it is colder. That space becomes an aisle.

A Simple Grow Light Setup

In July and August we are busy seeding beets, head lettuce, green onions, kale, and other crops for the fall and winter months. At this time of year the propagation greenhouse is too hot for us to consistently germinate crops well. So we move our operation to the basement of our house where the temperature is a stable 60 to 65°F. We also use the basement for starting

Peppers under T8 grow lights, an economical way to start seeds when a propagation house is not available, or during times of the year when it is difficult to control the greenhouse environment. The lights are raised for this picture, but normally they are kept 3 to 4 inches above plants. We use a timer that turns the lights off for eight hours each night.

tomatoes, peppers, turnips, radishes, bok choy, green onions, spinach, and head lettuce in January and early February.

Our setup consists of two heating mats, 12 inches by 11 feet, placed side by side on tables. We place rigid insulation under the mats to direct heat upward. Over the tables we hang 4-foot-long six-bulb fluorescent ballasts that house t-8 bulbs. We installed two full-spectrum grow lights in the middle of each ballast; the remaining four are standard daylight bulbs. This arrangement supplies more than enough full-spectrum light, and saves money, as daylight bulbs are much less expensive. We keep the lights 3 or 4 inches above plants to keep crops stocky. We keep a small fan running to stiffen stems.

We use clear humidity domes over trays we are germinating in order to help hold moisture in, which increases germination. After seeds pop, the domes come off. A challenge with domes is heat buildup from the lights. We keep a soil thermometer under the domes to keep tabs on temperature. If it is too hot, we prop the domes open for ventilation. Another technique is to use a small fan to blow air between the dome and the lights, scattering the heat. We also sometimes use 7-inch-deep propagation domes with ventilating tabs, rather than standard 2-inch domes, to provide a larger air space.

Potting Up

We transfer peppers and tomatoes from 10-row germination flats to 50-cell flats, and then to 4-inch pots, before transplanting. This sequence of steps is called *potting up*. The bigger pots give crops room to expand their roots as they grow. This also spreads the transplant window, giving more time to prepare soil or wait for ideal transplant weather. We have found that with high-quality compost-based potting mixes, no additional fertilizing—with fish emulsion, kelp, or minerals—is needed. We typically transplant these crops at between six and eight weeks of age, compared with four to six weeks for other crops in plastic flats. (The timing for paper pot crops, which I discuss in chapter 6, is shorter.)

A challenge with basement growing is watering without creating a mess. Our solution is to water the trays well at the time when we seed them. (We always seed flats in the propagation greenhouse even if we later move them to the basement.) We then rely mostly on bottom watering as the plants grow. If we must water from overhead, we use a watering can.

"Half Inch to Zero" Bottom Watering

To water starts, we rely as much as we can on bottom watering: setting flats with transplants into a shallow pool of water, which allows roots to take up water from below. We place our 10-by-20-inch flats into leak-proof 1.3-inch-deep support trays filled with ½ inch of water. To bottom-water larger paper pot chains (see chapter 6) we use cafeteria trays, although because they rely on air-pruning, we only bottom-water paper pots occasionally. (We are waiting for a manufacturer to take up the challenge of making affordable bottom-watering flats that fit the paper pot system, to replace the bulkier cafeteria trays.) Whatever the seedling system we are using, at least once daily we inspect seedlings and if we see dry soil on top, or if seedlings appear droopy because of thirst, we add another ½ inch of water to the support tray below.

It is important to give roots occasional access to air, which they need to grow well. If plants sit continuously in stagnant water, they do not grow well and become diseased. This is why we use the phrase *half inch to zero*: we let the water level reach zero before adding more. We use a 4-foot level to make sure our potting benches are perfectly level; otherwise flats won't water evenly.

The flat of kale in front is supported by a 1.3-inch, shallow, white leak-proof tray, which allows for bottom watering. Other flats sit in used cafeteria trays, which also work well. We use cafeteria trays with all paper pot flats because paper pots do not fit inside the leak-proof trays.

There are three advantages to this method:

1. Plant leaves do not get wet, which reduces the odds of foliar diseases such as gray mold.
2. Plants soak up the amount of water they need, when they need it. As they do in the ground, plants through their roots can sense water and, to a degree, self-regulate their water uptake.
3. For the grower, bottom watering saves time. With top watering, on a hot day we might need to water four times a day. With bottom watering, on the hottest days we need to refill just once, and many days we just check in. If we will be away for a night or two, we make sure all the trays are filled to at least ½ inch with water immediately before we leave.

A word of warning with bottom watering: it can be easy to oversaturate trays, especially those with small cells. If a potting soil is too wet, young seedlings can dampen off. For this reason we bottom-water little if at all during cloudy stretches, and we are extra careful not to oversaturate newly germinated plants. The best way to gauge saturation, besides observing the soil, is by lifting the top tray and sensing the weight. With practice you will be able to feel when flats need more water.

While using leak-proof trays is the simplest way to bottom-water, another method is to nail 2-by-2-inch lumber around the perimeter of a bench to create a short wall, and then cover the bench with EPDM, a rubber membrane used to line ponds, thus creating a swimming pool for trays. With this method, it is especially important for tables to be level. Another method is to bend sheet metal to hold water. Some growers leave the two ends of the sheet metal open, which lets water continuously flow through the trough. Water comes supplied through PVC pipes and is carried off by a gutter. Creative systems like these put water where it does the most good: at the roots.

CHAPTER 5

Transplanting by Hand

We need to think of motion that includes human wisdom as being something completely different from animal-like motion.

—TAIICHI OHNO

When we started farming, we direct-seeded nearly all of our crops into the ground rather than using transplants from a seedling greenhouse, because it was faster. When you can plant thousands of seeds in a matter of minutes, we thought, why bother with seedlings?

But this was shortsighted thinking. Seeds failed to germinate when the soil dried out in hot sun, or they became waterlogged after a rain. If they did germinate, weeds often smothered them. Much was left to chance; defect costs were high.

When we started to implement lean, we realized that to reach our goal of 100 percent success—of every seed turning into cash—we could not tolerate those costs. By using transplants, we can be certain seeds will germinate and that crops will go into the ground weeks ahead of the appearance of weeds. Also, with transplanting we can space crops perfectly, saving costly time spent thinning. While we still direct-seed a few crops when the weather is just right, most of the time we now rely on transplanting. We have developed methods to transplant turnips, radishes, baby greens, green beans, and other crops that many farmers only direct-seed.

The key to successful transplanting is efficient motion. Transplanting requires a lot of movements: bending over, poking holes in the ground, and watering plants in. *This work might be tedious, but it should not be mindless.* It is critical when transplanting to analyze every step and find ways to simplify and eliminate waste. In this chapter I will discuss how we leaned up hand transplanting. In the next I will show how we use a Japanese-designed paper pot transplanting system for even faster transplanting.

We transplant nearly everything we grow to increase the odds of a crop's success. When planted 10 inches apart, many crops fill in nicely and crowd out weeds.

Simpler Crop Spacing

In our quest for the simplest possible system, we transplant almost all of our crops either 6 inches or 10 inches apart. To mark rows we slip 6-inch PEX plumbing tubes over bed-rake tines, leaving six tines (about 10 inches) between tubes. Walking one end of the bed to the other, we mark the beds by dragging the tubes along the ground, and then we mark rows across the bed, creating a grid. For 10-inch spacing, plugs go in at every crossing point; for 7-inch spacing a plug goes in at every crossing point plus in the middle between crossing points. For crops planted at just one or two rows per bed, we do not use this system. Instead we find it more efficient to mark 6-foot stakes that we lay on the ground beside us as we work, putting in plants at each marking.

Table 5.1. Recommended Spacing of Single- and Double-Row Crops

Crop	Number of Rows	Spacing (Inches)
Chard	2	9
Cucumber	1	12
Eggplant	2	18
Ginger	1	12
Kale	2	18
Peppers	2	18
Potatoes	1	12
Tomato, determinate	1	18
Tomato, indeterminate	1	9

Three PEX plumbing tubes inserted over the tines of a bed rake for marking rows. Note six open tines between tubes. This is our standard spacing.

Multiplant Plugs

While we usually use paper pots now for large-scale transplanting (see chapter 6), it is helpful to see our hand-transplanting method, too, which we use for smaller planting. For most hand-transplanted crops, we seed one seed per cell. However, we use multiseeding—putting several seeds in one cell—with crops that can tolerate closer spacings. This eliminates moves at transplanting time, cuts down on space in the propagation house, and allows us to transplant crops that otherwise are typically direct-seeded. Examples include the following:

Turnips. We seed Hakurei turnips into 72-cell flats and transplant them three rows to a 30-inch bed, 10 inches apart, after about four or five weeks, just after true leaf stage but before roots bind up in the cells. We drop a pinch of seeds, aiming for four to six, in each cell. That requires very little cultivating later, because the crop is far ahead of the weeds. The roots grow away from one another and eventually reach full size. At harvesttime we sometimes pull entire clusters; other times we choose the largest turnips from several clusters, leaving others to grow.

Radishes. We direct-seed radishes when conditions are just right, but for an early crop we seed them in 72-cell flats at four to six seeds per cell, and

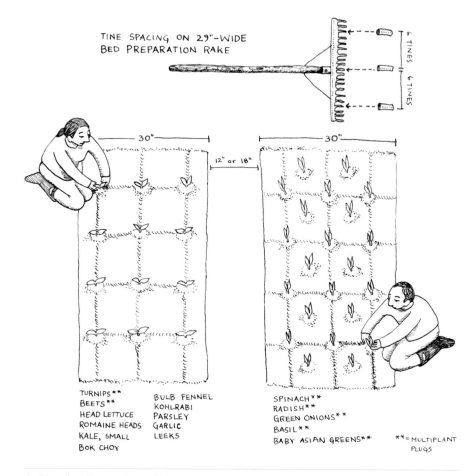

TINE SPACING ON 29"-WIDE
BED PREPARATION RAKE

6 TINES 6 TINES

30"

30"

12" or 18"

TURNIPS**   BULB FENNEL
BEETS**   KOHLRABI
HEAD LETTUCE   PARSLEY
ROMAINE HEADS   GARLIC
KALE, SMALL   LEEKS
BOK CHOY

SPINACH**
RADISH**
GREEN ONIONS**
BASIL**
BABY ASIAN GREENS**   ** = MULTIPLANT
                              PLUGS

A Simpler Transplant Chart. We are always in pursuit of simplicity on the other side of complexity. This chart simplifies hand-transplanting of crops planted three rows or more per bed. Ideally, we space crops closely enough to crowd out weeds but far enough apart to achieve full growth.

transplant them at five rows per bed, with 7 inches between plugs. The primary advantage with multi-seeded radishes is that we are able to supply radishes much earlier in the year. In January and February we let them grow for around three weeks before transplanting to a hoophouse or greenhouse, where they take off quickly.

Peas. For an early stand of peas we use 50-cell flats and drop four seeds in

Early peas seeded four to six seeds per cell in a 50-cell flat.

Turnips and radishes transplanted in clusters of four to six will grow away from one another.

Table 5.2. Recommended Plug Flat Size for Vegetable Crops

Number of Cells	Crop
128 at 1 seed per cell	Head lettuce, romaine, multileaf lettuce, bulb fennel, bok choy, parsley
72 at 1 seed per cell	Kale, Swiss chard, kohlrabi, broccoli, cauliflower, cabbage
72 at 4 to 6 seeds per cell	Radish, beets, turnips, spinach, Asian greens
50 at 1 seed per cell	Watermelon, cucumbers, zucchini
50 at 4 seeds per cell	Peas, green beans
10-row seedling flat	Tomatoes, peppers, eggplant, basil
Open 2-inch flats at ⅛-inch spacing, 5 rows	Alliums (onions, green onions, shallots)

Multiseeded spinach in 72-cell flat ready for transplanting.

each divot. We germinate them in a 68°F germination chamber. We transplant when the peas are 3 to 4 inches tall. We transplant them in one row for ease of trellising at a spacing of 4 to 6 inches between plugs.

Mizuna. For faster mizuna in the midwinter greenhouse, we transplant plugs rather than direct-seeding. The mizuna grows for two to three weeks under grow lights in our basement before being transplanted out. We use 72-cell flats with four to six seeds per cell. We space them 7 inches apart, using five rows per 30-inch bed.

Spinach. As with mizuna, we prefer to transplant midwinter spinach to shorten seed-to-harvest time. We use 72-cell flats and seed four to six seeds in each cell, and transplant in five rows, leaving 7 inches between plants.

Transplanting Technique

Seedlings require tender care if they are to transition successfully from a plug flat to the ground. Proper technique is a must.

BEST TIME TO TRANSPLANT

Wait for cloudy weather and if possible, avoid transplanting when the temperature is above 80°F. Also, avoid planting in high winds, which add stress and easily kill tender transplants. If we must transplant on a windy day we cover the plants immediately with an agricultural row cover. Ideally, we transplant in the late afternoon as the sun starts to recede. This gives plants a long window free of direct sunlight in which to recover.

Many growers acclimate transplants by setting them outside the greenhouse a day or two before transplanting. The process is called *hardening off*. In our experience, this extra step is rarely needed, especially if we already provide plenty of fresh air in the seedling greenhouse and do our transplanting in calm weather.

WATERING TO PREPARE FOR TRANSPLANTING

Rather than water crops after they are transplanted, we prefer to pre-moisten both soil and plug flats. First, we soak plug flats by setting them in trays filled with water. We aim to soak flats for at least 30 minutes before transplanting. Second, we pre-wet soil, thoroughly saturating it but making sure it is still workable, with overhead irrigation or drip tape so that we are always planting into wet ground. This give roots immediate water and greatly increases the odds of success compared with transplanting into dry ground. Pre-watering also requires less effort than watering transplants as they go in.

Flats of kale pre-soaking before being transplanted. The soil was also pre-moistened, the best way to ensure a successful transplant.

DECIDING WHERE TO PUT TRANSPLANTS

As mentioned earlier, we do not use field maps but rather plant wherever there is an opening. However, we consider a few factors. First, we avoid transplanting many crops next to tall plants like mature cherry tomatoes, unless the transplant can tolerate shade (like head lettuce). Second, for crops that require frequent harvesting—such as zucchini and cucumbers, harvested every other day—we look for open spaces close to our processing area, where we frequently walk. Third, to promote diversity, we try not to transplant the same crop in the same place twice. However, we have found that light feeders like lettuce can usually be double-planted without harm.

PLANTING INTO BARE GROUND VERSUS MULCH

We use a two-month rule: crops that will be in the ground about two months or more—leeks, onions, full-size kale, cucumbers, and rhubarb, for example—go into a type of mulch, usually plastic or ground cover. Otherwise we plant into bare ground. In our experience, crops harvested in less than two months do not warrant the expense of mulch. It is easy enough to cultivate them by hand, if needed. For longer-season crops, the mulch pays off. An exception to this rule is shallots. Although they are a long-season crop, we plant them so closely together that we don't need or have room for plastic or other ground cover such as landscaping fabric.

In general, we prefer planting into landscaping fabric rather than plastic because we can reuse the fabric. We burn holes into fabric with a butane torch. We have used some sheets for more than seven years. We pin the fabric to the ground with 8-by-1-inch heavy wire sod staples. A 4-foot-wide sheet of landscaping fabric works well for us since it covers the 30-inch bed top as well as the aisles. On larger farms disposable plastic mulch probably makes more sense than landscape fabric, since it is simply not practical to store 20 acres' worth of landscape fabric from year to year.

With the exception of garlic, we do not use straw, grass clippings, or other organic material. These require too much effort to spread, promote leaf diseases, and make cultivating difficult.

HOW TO PULL PLUGS FROM FLATS

To unload flats fast, wait to transplant until roots fill the cell and hold the potting mix in it together. Otherwise roots become unbound as you release the plug from the flat. Soaking trays also helps potting mix and roots to stay together.

Mulch Is *Muda*

Applying any type of mulch product—organic mulches such as straw or synthetic mulches such as plastic—is a form of *muda*, an activity that does not add value. Mulches eat up time to store and lug about, and they cost money. They might be necessary for the survival of a crop, but they do not directly add value. Consequently we are always looking to minimize their use. As our weed pressure diminishes every year—because of our weed management strategies—we rely on mulches less and less, and we envision eliminating them altogether, freeing up more time for seeding, harvesting, and other value-adding tasks.

To help release plugs from the cells we pinch them aggressively from the outside of the flat while gently pulling up on the plant. If plugs remain stuck, we insert a blunt-tipped rod or stick through the hole in the bottom of the flat. We often like to work in pairs at this stage, with one person releasing plugs and laying them on the ground while another transplants.

Usually our ground is loose enough that we do not use trowels. Instead, we create an opening with our hands, drop in a plug, and pull soil back around the plant. If soil is too firm for handwork, we create holes using a Clarington Forge dibbler, pressing its sharp steel tip into the ground. For most transplants we pull soil just over the top of the plug and press soil lightly around newly transplanted plugs, ensuring good contact between roots and new soil. Pressing too hard damages a plant and compacts the soil, making it difficult for oxygen to reach the roots. An exception to this practice is leeks. We transplant them into a deep hole, leaving just 1 or 2 inches of leaf above the soil. The rest of our alliums—shallots, onions, and green onions—we transplant shallow, just deep enough to hold the plant upright. For ease of handling, we cut off allium roots to a length of ½ inch just before transplanting.

Our favorite tool for hand transplanting: a Clarington Forge dibbler.

79

HOW TO USE ROW COVERS

We use row covers on many transplanted crops to protect them from frost and wind as well as insects. We skip the covers if we are transplanting on calm days with stable temperatures in the forecast. In most cases we use AG-30 (30 mm) from Agribon, a midweight cover that is strong enough to handle without ripping, while still allowing adequate light through. We use just one width—83 inches—which one person can easily handle.

As a general rule, at temperatures from 20°F to 30°F we use one layer of AG-30; from 10°F to 20°F we use two layers; and below 10°F (in the winter) we use three layers, or one layer of AG-50. What works varies, and on your own farm it is worth experimenting with the height of your covers, the weight, and the number of covers needed to protect specific crops.

We store row covers as "firsts" or "seconds": firsts are like new, with no holes, best for keeping out insects; seconds might have a few holes but still can serve for frost protection. With this practice we can use covers for up to four seasons before replacing. We store row covers in our unheated storage room, up in the air to reduce the odds that mice will make their nests in them. We also keep rodent bait stations nearby.

Contrary to common practice, we rarely suspend covers over wire wickets or hoops, but rather float them directly onto our crops. (The

Row covers offer wind protection over newly transplanted crops. To simplify our work, we rarely suspend row covers over hoops.

exception is alliums. We hoop those so the covers do not bend the plants' tips.) We do see occasional damage when a cover freezes to greens, but the damage is minor, not warranting the extra effort of employing wickets or hoops. The practice of floating rather than suspending covers not only saves time, but also traps more heat. At lower temperatures, an inner cover is essentially a thin layer of insulation trapping heat released from the soil. The less air to keep warm, the better.

To pin covers to the ground, we use polypropylene bags filled with sand, placing a bag every two paces—about 6 feet apart. The number of bags required depends on the amount of wind on your farm. We store the bags against greenhouse endwalls so they are easy to access when we need them. Though simple, practices like this save thousands of steps through the course of a season.

While we appreciate the benefits of row covers—we would not want to be without them when nighttime temperatures unexpectedly dip—we see managing them, like managing mulches, as a form of *muda*. Hence we are increasingly unenthusiastic about using them. For example, in two greenhouses we turn heaters up to 28°F, in part to avoid having to sling around row covers all winter (see chapter 13). We also sometimes forgo frost-prone plantings, choosing to wait a week or two in order to avoid using row covers. These decisions make our farming easier to do every year.

The paper pot transplant method dramatically reduces our transplanting time. Note that this tray of green beans is supported by a cafeteria tray, which we use to bottom-water before transplanting.

CHAPTER 6

Paper Pot Magic

*Progress cannot be generated when we are satisfied
with existing solutions.*

—TAIICHI OHNO

When our second son, Leander, was born in February 2016, Rachel and I decided it was time to farm less and devote more time to our family. Our goal was for me to reduce farming time to 35 hours or less per week and for her to care for our children as close to full-time as possible.

Our business was going well. We had a devoted following of CSA customers, and farmers' market and restaurant sales were strong. How could we cut back? Enter the paper pot transplanter, a tool designed to quickly transplant cells made out of paper. The two-wheeled tool is quiet, with no engine; you maneuver it by pulling it across the field. It cuts a furrow, slides paper cells into the ground, and packs the seedlings in, covering the root ball. With this tool, which we acquired in 2015, we were able to keep sales steady while dramatically cutting back work. While the system has been used for decades in Japan, it is relatively new to growers in North America. (For details on suppliers and specific accessories we use, see "Paper Pot Transplant Systems" in appendix 1.)

The paper pot system offers a number of advantages over plug flats and soil blocks. First is speed. With the paper pot gravity seeder, the seeding system that comes with the transplanter, I can seed a tray of 264 paper chain cells in less than 10 minutes, at least four times as fast as with other systems. Second, the compact trays occupy comparatively little room. Since switching systems, we have decreased the size of our propagation area by almost half, a significant savings because the propagation area—with its heaters, automated louvers, benches, and other infrastructure—is the most expensive space on our farm. Paper pots save time in the field, too. I can transplant those 264 plugs in about 40 seconds, from a comfortable upright position.

Paper Pot Transplanter Tips

1. *The tool is best suited to larger plantings.* We use the tool when transplanting more than 250 plugs at a time. For smaller plantings we still transplant by hand, using plug flats.

2. *The tool is not recommended for large seeds and wide spacings.* We do not use paper chains for large-seed crops such as squash and watermelon because those crops need bigger cells than the

Baby lettuce, spinach, Tokyo bekana, head lettuce, turnips, and mizuna ready for early-spring transplanting. It is best to transplant crops just after a true leaf appears.

While the paper pot system represents an investment of money and requires practice to master, the tool paid off quickly in saved labor, increased crop quality, and the ability to extend the growing season with transplanted crops. Here is how the system works.

How to Seed the Paper Chains

The paper chains arrive in the shape of a compressed honeycomb, which is opened with steel rods, inserted over a frame, and then dropped into a sturdy support tray, creating a flat of 264 cells, approximately 1.3 inches paper pots provide. We also do not use the system on crops spaced far apart in the field, like tomatoes and peppers.

3. *Multiplant cells work well here, too.* We use the multiplant cells with paper pots, as we do with plastic trays. Some multiseeded crops, like lettuce for salad mix, we seed into paper chains by hand rather than with the gravity seeder, because of their awkward shape.

4. *Think of it as "direct-seeding sprouted seeds" as opposed to transplanting.* Paper pot cells must be transplanted young—often at two to three weeks of age—rather than waiting for six or seven weeks, as with most seedlings in plug flats. In practice, with most crops in the paper pot system, we aim to transplant when just one or two true leaves form. It would be impractical to hand-transplant such young crops by hand. However, the paper pot transplanter handles the young transplants with little disturbance.

5. *Use all the same best practices as with transplanting by hand.* Pre-soak seedlings before transplanting, avoid transplanting during hot afternoons,

and use row covers if needed to hold off wind and maintain soil moisture. Using the transplanter saves time, but it doesn't mean you can skip these steps.

6. *The transplanter requires loose soil.* Growers with overly rocky or chunky ground might be frustrated by the paper pot system. To ensure smooth operation, we almost always quickly till our beds before transplanting with a tiller or rotovator, and pull out rocks and sticks with a bed rake.

When should you invest in a paper pot system? I recommend trying the paper pot system after you have mastered propagation with plug flats, and after you feel confident in transplanting basics. The paper chains cost between about $2 and $4 each, depending on length, meaning that each tray that fails is a cost. But once you've mastered the system, just one or two plants will pay for an entire chain.

Volume is also a factor. If you spend more than a few hours each week transplanting, I recommend giving it a try. Less than that and the investment might not pay off.

wide by 1.3 inches deep. The frame holds the paper chain open while the cells are filled with potting mix and seeded. After that, the paper chain can be released from the frame. The size of the support tray is approximately 12 by 24 inches, so paper chain trays are not interchangeable with standard 10-by-20-inch trays. The paper cells are open at the bottom, and the support trays are perforated, allowing roots to air-prune. They stop growing once they reach the bottom rather than becoming twisted and "root-bound" as occurs in plastic trays.

Several chain sizes are available, and the spacing of the plugs as they unravel varies. We use chains with 2-, 4-, and 6-inch spacing. The 2-inch chain unravels to 46 feet, the 4-inch chain to 88.6 feet, and the 6-inch chain

Fill the chains with standard potting mix, as with plastic plug flats. We take care to press extra potting mix around the perimeter cells, which don't fill as easily as cells in the middle. Photo courtesy of David Johnson Photography.

Paper chains arrive compressed and are stretched out over a frame.

A plexiglass dibbler creates 264 divots. Photo courtesy of David Johnson Photography.

to 131 feet. By seeding in a special pattern (see photo series on page 87), spacing can be doubled, allowing for 8-inch and 12-inch spacing options. A dab of adhesive holds the cells together until transplanting time. If you are certified organic make sure your certifier allows the glue for organic use. Most, but not all, certifiers now approve it.

We use standard purchased potting mix for the paper chains and fill them to the same density as plastic trays. We do take extra care to pack the outer rows, as they never seem to completely fill up on the first try. We then use a plexiglass dibbler to poke 264 holes into the potting mix.

Next we seed with the two-plate plexiglass gravity seeder, a tool that quickly seeds an entire tray with one motion. Unlike vacuum seeders, the device is non-electric and features just three parts: a plastic frame plus the two ⅛-inch-thick plates. The bottom plate is fixed to the frame and is drilled with 264 holes that line up with the cells in the paper chains. The interchangeable top plate is initially offset and is drilled with another 264 smaller holes. We use three top plates with varying hole sizes to accommodate

Jiggle seeds around the two-plate gravity seeder until each hole is filled. Pelleted seeds work best, though we use non-pelleted seeds as well. If you experience static electricity, dampen the plates very slightly with a cloth.

With holes filled, simply press the top plate to align the holes, and seeds will drop through. Usually around 5 percent of seeds get stuck; we use a pen to dislodge them.

We use painter's tape to mask off this pattern if we want to fill every other cell, which doubles in-row spacing.

We cover with light vermiculite, spray with water, and set the trays into a germination chamber.

We grow paper pot crops alongside others in our propagation greenhouse. One advantage of the system is that thousands of transplants occupy relatively little space, saving heat. Photo courtesy of David Johnson Photography.

Table 6.1. Ben's Paper Pot Cheat Sheet

Crop	Seeder Top Plate Size*	Paper Chain Size	Seeds per Cell	Rows per 30-Inch Bed
Asian greens, baby	Hand-seeded	2-inch	4 to 6	5
Basil	Hand-seeded	4- or 6-inch	3 to 4	2
Beets	4 mm	2-inch for baby beets, 4-inch for full-size	1 to 2; pellets not needed	3
Carrots	4 mm	2-inch double-seeded	Use pelleted seed	5
Edamame	Hand-seeded	2-inch	1	2
Fennel, bulb	Hand-seeded	4-inch and seed every other cell for 8 inches between plants	1	3
Green beans	Hand-seeded	2-inch	1	2
Green onions	4 mm	2-inch	1 to 3	3 or 4, depending on size
Head lettuce	4 mm	6-inch, or seed every other cell for 12 inches between plants, depending on target size	1; use pelleted seed only	3 or 4, depending on size
Lettuce, baby	Hand-seeded	2-inch	4 to 6	5
Peas	Hand-seeded	2-inch	1	2
Radishes	2.8 mm	2-inch	1 to 2	5
Romaine heads	4 mm	4-inch and seed every other cell for 8 inches between plants	1	4
Shallots	2.8 mm	2- or 4-inch, depending on size desired	1 to 2	3
Spinach	5 mm	2-inch for baby leaf, or 4-inch for midsize leaf	3 to 4	5
Turnips	Hand-seeded	6-inch	3 to 4	3 or 4, depending on size desired

* Seed sizes vary, so you might need to experiment with different top plate sizes. These sizes work most often for us.

different sizes of seed (see table 6.1). We seed very small seeds by hand, not with the gravity seeder, because they fall between the plates.

To operate, dump a generous amount of seeds into the seeder, shake it back and forth until each seed finds a hole, and then line up the plates. The seeds will drop through. After seeding, we sprinkle fine vermiculite over the tray and lightly water, as with plastic trays. All 264 cells do not need to be filled with one type of seed. We frequently hand-seed multiple varieties in one chain, as long as they share the same spacing.

The paper pot system does not require the two-plate gravity seeder. To save cost, you can seed all seeds into paper pots by hand. Also, you can use a vacuum seeder with seed plates sized to fit paper pot chains (see "Seeders" in appendix 1).

How to Use the Paper Pot Transplanter

In Japan the name of the transplanter, *hipparu-kun*, translates to "little pulling buddy." The name is apt. Your primary action is to pull the tool as you walk backward. Your "buddy" does everything else, cutting a furrow, unraveling your transplants, and packing soil around them.

A furrower blade is fixed to the underside of the transplanter. The soil should be loose so that you can pull the planter easily. As mentioned, we usually quickly till with our BCS 710 ahead of the transplanter. We also make sure the soil is free of large sticks or clumps, as these can jam the furrower. We remove them by hand or with the bed rake. The furrower can be adjusted up or down, and it might take a pass or two to figure out the right height for your conditions:

The paper pot transplanter fits nicely between rows of tomatoes, where we transplant crops that will be harvested by the time tomatoes are ready for more room.

too deep and you will bury plants, too shallow and you will leave them on the ground. We set the furrower as deep as it will go and raise the furrowing depth by adjusting the amount of pressure we put on the handle. For ergonomic use, match the transplanter to your height by adjusting the handle length.

To load paper chains onto the transplanter, slip a steel bridge between the paper chain and the floor of the support tray. Before we unravel the chain, we pull the transplanter a few feet to establish the start of a furrow. Then we unravel a few feet of the chain by hand and stake the first cell to the ground with a screwdriver.

We can comfortably transplant up to five rows per bed, if needed. We start transplanting down the middle of a 30-inch bed. For this pass it is easiest to straddle the bed. For three-row crops (planted three rows per bed), we then do a pass just inside either edge of the bed, walking with both feet in one aisle. For five-row crops, we perform an additional pass between the middle and outer rows.

This photo shows direct-seeded Nelson carrots, which grew nearly perfectly straight.

As a transplant crop, carrots must go in the ground very young. These are just a few days post-germination.

These carrots, also Nelson, were transplanted and grown at the same time as carrots above. Eighty percent were marketable as firsts. Transplanted carrots are as tasty as any others, but never as straight.

Standard Uses with the Paper Pot Transplanter

Hakurei turnips. Turnips seeded three to four seeds per paper chain cell will grow away from one another and still size up well. We grow them at three rows per 30-inch bed, 6 inches apart in the rows.

Romaine and head lettuce. We grow romaine and head lettuce exclusively with the paper pot method. We grow romaine 8 inches apart by seeding every other cell in a 4-inch chain. We use a 6-inch spacing for multileaf lettuces and small heads, and we seed every other cell for large heads.

Green onions and shallots. Green onions and shallots, spaced perfectly with the paper pot method, will require no thinning.

Radishes. We still usually direct-seed radishes, but for an extra-early crop we grow them in paper pots, with one or two seeds per cell. Paper pot radishes are not as uniform in shape, but because of their earliness, they still sell well.

Beets. Direct-seeded beets are finicky germinators. With the paper pot system we germinate them in controlled conditions and eliminate thinning. We space them either 2 or 4 inches apart with three rows per 30-inch bed.

The paper chains unravel easily, creating ideal spacing.

We start by creating a 3-foot-long furrow running from the front wheels to the middle of the transplanter. Then we use a screwdriver to stake the first cell. The bright handle on the screwdriver helps remind us to pick it up. Photo courtesy of David Johnson Photography.

We transplant by hand if we have just one chain to transplant. This is faster than getting out the transplanter. Simply dig a shallow furrow the length of your chain, unravel the chain, drop it in the trench, and fill dirt around it with a hoe or by hand.

We continue to experiment with different crops in the paper pot seeder. One alluring crop is carrots. With their slow germination and vulnerability to weeds, we are motivated to find a way to transplant them. So far, results with paper pot carrots have been mixed. When seeded very young, just a few days after germinating, transplanted carrots do reach full size. However, because their taproots grow quickly, the paper pot carrots we have harvested are never as straight as we'd like. At best, 80 percent are straight enough to market as firsts. They rest we sell as juicing carrots.

Pull back on the handle and walk backward as the "pulling buddy" spaces transplants and packs soil around them. We frequently go back and touch up a few areas by hand, but with practice the transplanter covers soil around seedlings well. To alter planting depth, we adjust pressure on the handle. Photo courtesy of David Johnson Photography.

Tricks and Special Uses
with the Paper Pot Transplanter

Winter greenhouse greens. When direct-seeded, midwinter baby greens in a cold frame or minimally heated greenhouse can take 12 to 16 weeks from seed to harvest. Transplanted greens can be ready in eight weeks. This photo shows Defender romaine, Dane lettuce, and red mizuna transplanted at five rows per 30-inch bed.

Fall spinach. Spinach ready for transplanting. In northern Indiana the fall season is often too short to support a crop of outdoor spinach because soil temperatures are frequently too hot for reliable germination. So we seed spinach in a cool basement and then transplant it.

Field greens. Baby lettuces were planted at the same time, in paper pots (background) and direct-seeded (foreground). Transplanting yields an early crop.

Greenhouse beans. These green beans were seeded in March and transplanted the first week of April, giving us a May crop weeks ahead of normal.

Green beans. We transplant beans because we can be assured they will germinate, we can eliminate weeding, and we can extend their season. We transplant two rows per 30-inch bed. We seed a small amount every few weeks for steady harvests.

CHAPTER 7

Four Lean Tips for Direct-Seeding

Convince me that you have a seed . . .
and I am prepared to expect wonders.

—Henry David Thoreau

A s the previous two chapters have shown, we have designed systems for transplanting all of our crops at least part of the time. Transplanting has cut defect waste to near zero. With less weeding and thinning to do, transplants also shave off enormous motion waste.

A greenhouse seeded with direct-seeded crops. As a general rule, baby greens are more efficiently direct-seeded than transplanted, though you should choose your most weed-free soil when seeding them.

Still, on occasion we direct-seed certain crops, for example:

- Baby greens in spring and early fall
- Radishes in midspring, after soils have warmed, and early fall
- Carrots in spring
- Peas in midspring, after soils have warmed

With good technique and timing, these crops in these seasons will germinate reliably and quickly when direct-seeded. On the proper ground they will outcompete weeds with minimal cultivation. With the right seeder, plants emerge the proper distance apart, saving time thinning. With these plants, the extra costs of transplanting—seeding into trays, tending starts in the greenhouse, and setting out plants—do not pay off.

On your own farm think carefully through the costs and benefits of transplanting versus direct-seeding with each crop in each season. When you choose to direct-seed, here are my tips.

1. Use a precision seeder.
2. Seed into slightly moist ground.
3. Keep soil surface moist until the seed germinates.
4. Space right.

1. Use a Precision Seeder

I suggest trying out different seeder models before buying, since the best seeder is often a matter of personal choice. As long you are comfortable with a seeder's operation, and understand the soil conditions required for its use, any number can serve a market farm well.

Before settling on a Jang JP-1 we used an EarthWay seeder, the Johnny's six-row seeder, a four-row pinpoint seeder, and an antique Planet Jr. seeder, among others. All work well in specific circumstances. The EarthWay does a good job of drilling raw carrot and lettuce seeds. The six- and four-row seeders work well in seeding baby greens, though they require near-perfect soil conditions. The Planet Jr. works well for larger seeds and in rough ground. For several years we kept a variety of seeders on hand.

In our quest for simplicity, however, we now use just the Jang JP-1. The Jang performed more reliably than others we tried in a wider array of conditions and with almost any type of crop, as long we chose the correct roller size.

The Jang works by dropping seeds from a hopper onto a 2⅜-inch roller with divots drilled into it. As the roller spins, seeds fall into the divots, resulting in near-perfect singulation, separation of seeds from one another. The hopper is made of clear plexiglass so a worker can easily see when the roller is dropping seeds. The hopper can be removed for easy filling and emptying; many other seeders require awkwardly lifting up the entire seeder to unload it. The roller is propelled by a chain rather than the more commonly used belt; the chain ensures no slipping. A plated steel furrower opens up the soil. It is possible to "gang" several Jang seeders together to seed several rows at once, though on a small scale this isn't necessary (see table 7.1).

A less costly option I recommend for those on a budget is the Earth-Way. Many experienced farmers still exclusively use EarthWay seeders. Through made of lighter materials, EarthWay seeders can last for years if stored properly.

On very small beds—tiny plantings of radishes in the spring, or carrots in the winter greenhouse—we still seed by hand. This gives us precise spacing without the hassle of getting out a seeder.

Table 7.1. Recommended Roller Sizes for the Jang JP-1 Seeder

Crop	Roller Model*	Brush (Size of Aperture)	Spacing Settings (Inches)**	Rows per 30-Inch Bed
Arugula/baby kale	X-24	Low	½	7
Baby Asian greens	F-24	Low	2½	7
Baby lettuce	F-24	Low	½ or 2½, depending on leaf size desired	7
Carrots	LJ-24, 13 mm pellets. Double-seed for higher density.	High	½	5
Cilantro	LJ-24	High	½	5
Radish (Rover variety)	X-24	Medium	½	5
Spinach	LJ-24	Medium-high	½	7

*You can slightly adjust for larger seeds by raising the brush inside the hopper. A "low" setting means that only the smallest slit of light can be seen between the roller and brush. "High" means just under the height of the seed being used. Because seed sizes vary—even within one variety—these roller sizes might not work with all crop varieties. When trying new varieties, test a few roller sizes and brush settings the first time you plant.

**Printed on the seeder is a chart indicating spacing options depending on the sprockets being used. For simplicity, we stick to two spacings, ½ inch and 2½ inch, because these can be set by pulling off and reversing the chain and sprockets, eliminating the need to find and replace sprockets. I consulted with Jeremy Mueller at Excelsior Farm, Pleasant Hill, Oregon, to develop this table.

Thirteen-millimeter pelleted carrots seeded with an LJ-24 roller. The Jang seeder singulates seeds better than other seeders we have used, reducing time thinning crops seeded too closely together. Because pellets transfer moisture, they do not store well. I recommend ordering the amount you need at seeding time rather than storing them.

Among the reasons we chose a Jang over other seeders is its reliability. Almost anyone can be trained to use it.

2. Seed into Slightly Moist Ground

Seeds germinate better in slightly damp ground because moisture in the soil immediately coats the seed. If ground is bone-dry at the time of direct-seeding, we water and till to evenly distribute moisture ahead of seeding time. As with transplanting, the goal is to create a bed that is a bit moist but not too soggy to work. The Jang seeder features a heavy rear wheel padded with foam—called a *press wheel*—that compresses soil against the seed, which also increases moisture against the seed.

3. Keep Soil Surface Moist until the Seed Germinates

After seeding, it is critical to maintain soil moisture until seeds germinate. This means keeping an eye on the soil surface and making sure it stays wet. While this is most critical for slow germinators such as carrots, we follow the practice for all direct-seeded crops.

My favorite technique, alongside pre-irrigating soil, is to water newly seeded plots by hand, using a high-pressure hose with a watering wand. The Jang seeder leaves a small valley that collects this water and lets it soak in over time. This technique works especially well with summer-sown carrots. On larger seedings we forgo the watering wand and use overhead irrigation and timers.

After a soaking, I cover new stands with a midweight row cover, AG-30, to hold moisture and slightly warm the soil. In hot weather I use as many as three layers of row cover over cool-loving crops such as spinach. This reflects heat up and away, cooling the soil while retaining moisture.

Even with row covers and irrigation timers, your own observation is

After seeding crops, we water them well by hand or with overhead sprinklers on timers. The heavy rear wheel of the Jang seeder creates a trough that catches water nicely. The key is not to let the soil surface dry out.

the key to success. Almost every day I walk the farm, peeking under covers to check if the soil surface is moist in newly seeded beds. As needed I touch up dry spots by hand, adjust timers, remove row covers for watering, or add row covers to hold water in—whatever it takes to reduce the defect of poor germination. A bit of nerdiness, in this case, cuts out waste.

4. Space Right

Finally, successful direct-seeding requires the right spacing. Widely spaced rows waste valuable ground, causing you to weed and walk more than needed. Ultra-dense seeding, packing as many seeds as possible into a small area, can also lead to waste. In our early experimenting with dense seeding, we sometimes seeded as many as 24 rows of greens and 12 rows of carrots and radishes onto 30-inch beds. While these crops did shade weeds more

For five- or seven-row spacing, we start with a pass down the middle, followed by a pass along each outer edge of the bed. We follow that with one or two passes between middle and outer rows.

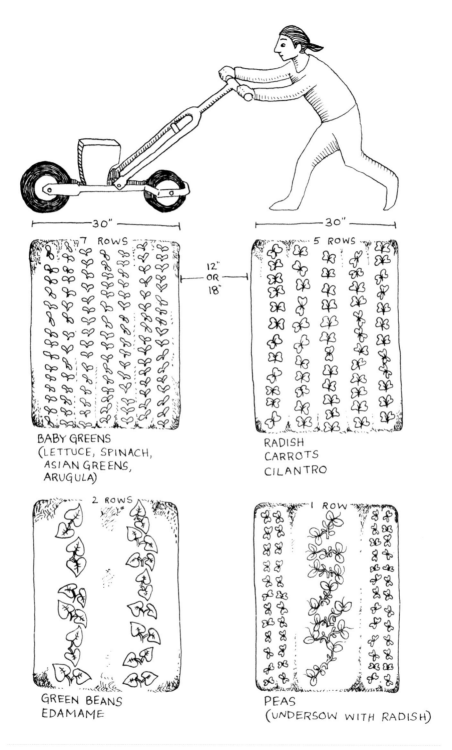

BABY GREENS
(LETTUCE, SPINACH,
ASIAN GREENS,
ARUGULA)

RADISH
CARROTS
CILANTRO

GREEN BEANS
EDAMAME

PEAS
(UNDERSOW WITH RADISH)

Spacing Direct-Seeded Crops

effectively, we found that molds loved to invade our crowded greens. Also, weeding was all but impossible and crops rarely grew to full size.

The best is to aim for balance. Space crops closely so that they shade weeds well and economize space, but give them enough room to fully mature and to allow for cultivating with a hoe (see illustration on page 101).

While proper spacing of plants pays off, this doesn't mean rows must be perfectly straight. Your customers pay for your crops, not straight rows, though overly curvy rows can make cultivating with hoes a challenge. To seed a row, we fix our eyes on a point far in the distance and, seeder in hand, walk toward that point. While rows are never perfectly straight with this method, this is much faster than fiddling with strings, stakes, and other row-marking gear. I learned this tip as a teenager mowing golf course greens at the local country club, where straight lines really did matter.

CHAPTER 8

Weed and Pest Control— without *Muda*

The most dangerous kind of waste
is the waste we don't recognize.

—Shigeo Shingo

When I visited the Aluminum Trailer Company at the start of our lean journey, I saw a semi truck pull into the parking lot and back up to a loading door. Workers came out and, with the driver, unloaded long pieces of tubular aluminum.

I walked inside the factory. There I saw these same pieces being moved to the start of long assembly lines, where there were sorted, lined up, cut to length, and labeled. Then the magic began. Welders, in mere minutes, with skill and a carefully conceived process, turned the stacks of tubes into high-value trailers, seemingly as fast as I could walk down the assembly line.

According to lean thinking, this work of the welders and others assembling parts is the essential work of the factory. Everything else—getting stock to the door, unloading the truck, pushing parts around the floor—is just motion, a form of *muda*. Even work in the factory offices—accounting and building websites and making photocopies—is *muda*, perhaps supporting value creation down on the line but not directly adding value.

Hassling with weeds and pests is classic *muda*, since it does not directly add value. For sure, many times there is no avoiding this work. It is part of organic farming. But when you yank weeds and spray bugs you steal time from seeding, washing, and selling—activities that bring in money. Rather than focus on *managing* weeds and pests after they have arrived, we focus on *preventing* them in the first place. Here are tips to lean up this work.

Five Steps to No Weeds

The most important weed-control method we use on our farm is to *transplant* everything we can, giving our crops a good head start on weeds. The next is to *"flip," or rotate, our beds between crops* as quickly as possible, keeping the farm full and never letting weeds take over. When possible, we till shallowly—or not all—when rotating, to avoid bringing buried weeds to the surface. As mentioned, we also *plant any crop occupying ground for two months or more into mulch, usually landscaping fabric.* These methods, covered in previous chapters, have reduced the bank of weed seeds in our soil to very low levels. Weeds are diverse and wily, however, and because all farms are unique, there is no magic bullet to eradicating them. Here are five more ways to keep weeds from gobbling up your time.

1. Direct-seed into your most weed-free ground.
2. Time direct seedings to avoid weeds.
3. Apply the June 1 rule.
4. Control the perimeters.
5. Never allow a weed to seed.

1. DIRECT-SEED INTO
YOUR MOST WEED-FREE GROUND

I scout for relatively weed-free ground all the time. Sometimes I even mark such areas with little flags. When it comes time to direct-seed, we choose these relatively weed-free beds. We save the weedier patches for transplant crops. Beds likely to harbor weeds include those where we let weeds go to seed, or that showed strong weed pressure in the past.

Here are examples of beds likely to harbor *few* buried weeds:

Beds following crops that clean a field. A powerful strategy to cut out weeding is to plant weed-sensitive crops like beets, shallots, and salad greens after a crop that has been managed to smother weeds. Here are three examples from our farm.

• *Beds following kale.* Outdoors we transplant kale on April 1 and don't pull it out until Thanksgiving or sometimes the following spring. We always plant kale into a plastic mulch or landscape fabric. This means that beds following kale have had weeds smothered for almost an

What's So Wrong with Weeds?

A common myth is that it is possible on organic operations to "farm with the weeds." In our experience this is a romantic notion that does not work. For certain, some amount of weeds can be tolerated. There is no reason to cultivate small weeds just ahead of a green bean or head lettuce harvest, for instance. Even tall weeds are acceptable at the end of a season, as long you turn them in before they go to seed.

However, the lower the weed population on your farm, the better. Weeds compete with your plants for nutrients. They slow down a crop's growth and diminish the size of harvestable crops. When taller than a crop, they cause plants to grow in their shadows. They also gobble a farmer's time, as soils with high weed seed populations demand much more energy to manage than soils free of weeds.

entire year. That creates a good place for direct-seeded carrots, salad mix, and other weed-sensitive crops.

- *Beds following garlic.* Like kale, our garlic is in the ground for a long time—from September or October until the following July—and it grows under straw mulch. Beds following garlic are ripe for direct-seeding because weeds in the soil have been smothered for almost a year.
- *Beds following carrots.* As carrots offer little leaf canopy to shade weeds on their own, we keep carrot beds relatively free of weeds by hand. When carrots are gone, we often follow with a dressing of compost and another weed-sensitive crop like baby salad greens.

Brand-new plots. After we till under sod or clover to establish a new plot, we let the plot mellow for a few weeks, letting remaining residues decompose, and then we seed direct-seeded crops. Often new plots offer a grace period of a season before weeds proliferate.

Plots following a weed-suppressing cover crop. Cover crops represent one of the best ways to "rescue" a weedy plot. Although we rarely use them, I describe their use here because when it comes to weed suppression it is best to keep many tools handy.

Good cover crop choices for summer weed suppression include buckwheat, cowpeas, sorghum sudangrass, and soybeans. These quickly grow a canopy that shades weeds. Clovers also smother weeds and grow in cooler

Sorghum sudangrass, which can be mowed more than once, adds incredible biomass to soils while suppressing weeds.

Red Ripper cowpeas eventually grow close together, suppressing weeds. Here I'm mowing them with a flail mower prior to planting a weed-sensitive crop.

We drilled these Red Ripper cowpeas, our favorite summer cover crop for weed suppression, with an EarthWay seeder, rows 10 inches apart. That allows for shallow cultivation while weeds are at the white thread stage.

We drilled this Soil Builder mix from Byron Seeds in the fall, with rows 10 inches apart, and we cultivated between the rows. This mix consists of KB Royal Annual Ryegrass, crimson clover, hairy vetch, and Nitro radish. The fibrous ryegrass roots will choke weeds by stealing their water while the other crops smother them from above.

A crop of buckwheat, the fastest-growing cover crop, in a summer greenhouse. The flowers signal it is time to mow.

weather. A good strategy is to combine these crops with an annual grass with a fibrous root system. The canopy crop shades the weeds while the root systems of the grasses steal their water and nutrients.

In our experience, cover crops accomplish weed suppression most successfully when seeded ultra-thick, sometimes 10 to 20 times the recommended density. We achieve thick density by drilling the seeds into the ground with our Jang seeder or with an EarthWay seeder with rows just 10 inches apart, and watering seeds well with overhead irrigation. Others use the dense seeding method as well. For weed suppression, Josh Volk at Slow Hand Farm in Oregon seeds a mix of 20 percent rye and 80 percent vetch at a rate of 400 pounds per acre, far denser than the recommended rate of 10 to 20 pounds per acre. For Volk, the extra money in seed pays off in the form of time saved cultivating later on.

With this approach it is important to monitor weeds that coexist with the cover crop. Even with dense seeding, we have found that a few weeds almost always find their way through a cover crop and threaten to send out a seed head before the cover crop fully matures. It takes only a few weeds to send out thousands of tiny seeds and contaminate ground for years.

Our solution is to shallowly cultivate between rows, killing weeds before they compete. If a few sneak through, we walk the plot and pull them or cut their heads with a machete. Another option is to mow the cover crop before weed seed heads mature. In some cases, such as with sorghum sudangrass, the cover crop will regrow, usually weed-free.

2. TIME DIRECT SEEDINGS TO AVOID WEEDS

We often time direct-seeding to avoid weed pressure. For example, we seed most of our carrots early, before lamb's-quarter and other summer weeds sprout. We even aim to seed our fall carrots, Bolero, by late May or early June in order to beat these summer weeds. The same applies to radishes, beans, edamame, and peas. It is best to get these in before the soil warms too much in order to put them ahead of weeds, or else to transplant them. Going into winter, we don't direct-seed after November, in part to avoid competing with cold-loving chickweed. Simply by adjusting *when* we seed, we shave off hours of time spent cultivating.

3. APPLY THE JUNE 1 RULE

Throughout much of the United States the month of June, with its mild temperatures and long days, supports more crop growth than any other month. Weeds love June, too, and grow rapidly. Every year we set a goal of

Flameweeding Pros and Cons

One technique to create a clean seedbed for slow-to-germinate crops like carrots and beets is to use a torch, or a gang of several torches welded together, to burn tiny weeds. The practice is called *flameweeding*. The weeds don't technically burn, but rather die through heat shock. A common flameweeding approach is to prepare beds, wait two weeks for weeds to grow, seed the crop straight into the weeds, and then flameweed just before crops germinate.

The major benefit is that you wipe out weed competitors just in time, as it were, giving your crop a boost. It is also faster than hand hoeing.

The practice comes with costs, however. Flameweeders for agricultural use, and the propane tanks required to fuel them, cost several hundred dollars. They might not pay off on farms growing a small amount of carrots and beets. Also, flameweeding is an extra management headache during planting time. Flameweeding involves risk, too. If you miss the window for flaming you have given weeds, not your crop, a leg up. Worse yet, you might burn your crop. Finally, flameweeders don't get all weeds, just the ones near the surface that germinate first. More usually follow.

Although we used flameweeders for a few growing seasons, we no longer do. Instead, our energy is better spent on good weed prevention practices, such as those described in this chapter. If you have persistent weeds in your carrot and beet patches, however, flameweeding might be worth a try.

having no weeds on property on June 1. It is a bit of an arbitrary target, and we do not always succeed in meeting it, but it gives us a tangible goal. In practice, this means that in the final weeks of May we plan to carve out time for cultivating weeds.

4. CONTROL THE PERIMETERS

Weeds have a way of creeping in from the edges of plots and up onto beds from pathways. We use a Honda FG 110 mini-tiller to cultivate the walkways between growing beds and the perimeters of our plots when needed during peak season. This leaves just a bit, the tops of beds, to hand-weed. The tool is lightweight and easy to use and can handle practically any size of weed. It also cuts a clean edge between growing plots and turf.

As an alternative, we sometimes use a Glaser wheel hoe with an 8- or 5-inch blade attachment for walkways and perimeters or between plants. The wheel hoe features two sturdy handles and a rubber wheel, making it

Creating a Stale Seedbed

If you have room to spare, one technique to control weeds is to create a stale bed—ground that is intentionally left open and then managed to reduce the population of weeds.

The easiest way to manage a stale bed, and the technique I would recommend starting with, is to shallowly and repeatedly cultivate. Wait for weeds to germinate, then turn them in. Any number of tools can accomplish this task: a walk-behind tractor, disk, spring tine cultivator, tine weeder, or rotovator. To avoid stirring up buried seeds, keep blades or tines shallow. To speed germination of buried weeds you can irrigate the ground or use row covers. You can use this technique over a period of months—for example, cultivating four or five times between May and August—or for a just a few weeks, for instance, to flush out weeds before a carrot seeding.

Another technique is to use clear sheets of plastic for this task, a practice called *solarization*. In sunny locations, clear plastic heats soil, encouraging deeply buried weeds to sprout and then "cook." An advantage with this method is that in addition to flushing weeds, the plastic holds down organic matter and protects the ground from excess rain. A challenge, however, is to make sure the weather is sufficiently hot. In cool weather the sheets just warm the soil and encourage weeds to grow. Opaque tarps can also be used for weed suppress-ing. They kill weeds better in cooler weather if given enough time—in some cases months.[1] We do not use plastic sheets for weed smothering because they require too much time and effort to manage, and because they take productive land out of commission. On our small footprint, our goal is for every square foot to hold a cash crop. Still, if weeds have taken over, or if you want to open up a new plot of ground without tilling, plastic sheets might be an option to try.

If you use sheets of plastic, I suggest these rules:

1. *Keep pieces small*. One person can manage a 50-by-50-foot sheet of plastic, but larger pieces catch a lot of wind and will require help. If your plot length is larger than the tarps, overlap two or more pieces (make sure the top piece faces downhill so water doesn't enter your plot).
2. *Place one sandbag every 4 feet around the edge of the sheets, at a minimum, to pin them.* In very wind-exposed areas no amount of sandbags will keep plastic sheets pinned.
3. *Only use plastic sheets where water can run off them.* Too much water sitting on plastic will be difficult to remove, especially if there is a dip in the middle. Ideally you can place tarps on ground that slopes a few degrees from end to end.

more efficient to operate than hand hoes, though less maneuverable. For most of our aisle and perimeter cultivating, however, we use the mini-tiller, as it requires less effort.

Cultivating with a lightweight Honda mini-tiller.

5. NEVER ALLOW A WEED TO SEED

Weeds are efficient reproducers. A female Canada thistle plant can produce more than 5,200 seeds in a season. Whatever it takes, we do our best to prevent weeds from getting to the reproductive stage.

The most efficient time to weed is when weeds are white thread stage—just out of the ground and before their soil-gripping root systems develop. At this stage weeds can be dislodged quickly with little effort. We reserve Monday afternoons for weeding, pulling out DeWit Half Moon and Johnny's Selected Seeds wire weeder hoes to disturb white thread weeds on bed tops and mini-tillers and wheel hoes for perimeters and pathways. Our goal is for two people to be able to weed the entire property in one afternoon. If we can't, then we've either put too much effort into production, or we've let weeds grow too big.

On rare occasions weeds might outgrow what we can handle with hoes. In this case we either yank them by hand, machete their tops to stop seed development, or till in a bed, crop and all, to stop weeds from seeding.

We use 4-inch wire weeder hoes for early cultivating of tiny weeds at the white thread stage. For very delicate jobs, such as pulling weeds growing next to carrots, we simply use our hands. The best time to weed is at the white thread stage.

Pests, Leaned Up

Nothing irks us more than finding aphids devouring arugula, hornworms munching tomatoes, or maggots taking chunks out of turnips. Pests not only eat crops, they eat profits. Customers are picky about vegetable appearance, and few will tolerate holes in their kale or a slug crawling out of a head of lettuce.

The chemicals sold to conventional farmers as insecticides might quickly kill pests, but they leave residues in the soil, destroying soil life, and on plants, where they find their way into our bodies. Unfortunately, their organic alternatives are often costly and, in our experience, rarely work as well as we hope.

Some people claim that pest problems are the fault of the grower—that with proper attention to soil health, crop selection, and crop rotation, plants naturally ward off pests. While these practices might help, our experience is that there is no avoiding certain pests on some crops. To grow vegetables

Beware Too Much N in the Greenhouse

A word of caution: too much high-N compost can invite aphids and other pests, especially in a winter greenhouse, as plants weep the excess N through their pores. For this reason, we generally avoid composting with high-N compost prior to seeding greenhouses for winter.

for a living requires at least a basic understanding of common pests, and a toolkit of low-cost and efficient strategies to stop them.

Our first, and most effective, line of defense is to simply to stop growing crops in seasons when pest pressure is highest. For example, because of insect pressure, we don't grow arugula or Asian greens during the hottest parts of the summer. Here are other techniques we use.

1. Row covers to exclude pests
2. Beneficial insects as preventive pest control
3. Sprays for pest control

1. ROW COVERS TO EXCLUDE PESTS

Our favorite way to deal with pests is to exclude them. After nearly every direct-seeding or transplanting session, we pull out AG-30 floating row cover. As stated earlier, the row covers aid in seed germination and guard transplants against wind. They also keep off bugs.

In most cases, to exclude pests, we simply need to pin down the covers with sandbags. If you suspect heavy pest infestations, I suggest covering the edges with soil, a more time-consuming technique but ultimately more effective. We use our newest covers, without holes, for this purpose. It is important to get covers on early. After a few pests arrive, it is often too late for complete exclusion, since the covers just trap pests inside.

Another option for covering plants is insect netting. The advantages of netting over row cover are its durability and breathability. It doesn't allow heat to build up as occurs with row cover. Some growers use netting on greenhouses to completely seclude insects from tunnels. At this point, we still prefer row covers because we already have the covers on hand for frost and wind protection. Also, insect netting is more expensive and for most of our crops likely would not pay off.

2. BENEFICIAL INSECTS AS PREVENTIVE PEST CONTROL

An area of newer development is the use of beneficial insects as an alternative method of pest control. Every year entomologists learn more about the relationships between predator and prey insects, and suppliers breed new species of beneficials for the marketplace. Popular insects for pest control include the fiercely named assassin bugs and soldier beetles, as well as earwigs, ladybugs, lacewings, and various small wasps, among many others.

Because of their cost, we use beneficials only in small doses, as a preventive measure rather than to combat an infestation. We have found that earliness is key. If we wait even a few weeks after pests arrive, we might be too late for total control. We have used purchased beneficial insects successfully against aphids on peppers, bok choy, and tomatoes, and we plan to continue experimenting with their use. Because availability changes frequently, consult with a supplier about which beneficials might best target your specific pests. We also grow nasturtiums every year in our pepper and tomato greenhouses to attract ladybugs and other local beneficial insects that naturally feed on pests. Also, we really enjoy huge sprawling nasturtium plants growing among plants in an otherwise straitlaced greenhouse.

3. SPRAYS FOR PEST CONTROL

When all else fails, we spray to control pests. Three key practices make the most out of spraying.

First, for the best chance of success, *spray early, at the first sign of infestation*. We scout fields daily, looking for the first hatching of Colorado potato beetles, the first butterflies on the kale, or the first hornworm on the tomatoes. To aid scouting, many growers use traps, tapes, or other devices to help spot pests, though we have not found such measures to be necessary on our scale. To ensure an accurate diagnosis, we keep reference books handy (our favorite is *Good Bug, Bad Bug*, by Jessica Walliser, St. Lynn's Press, 2011) and frequently perform Google image searches from the field. For a pest whose identity we are unsure of, I text an image to our local extension agent, who offers a diagnosis and a remedy.

Second, *avoid spraying in direct sunlight*; sunlight can break down many insecticides. My preference is to spray on cloudy days. If it is sunny, I spray early in the morning or, better yet, in the evening, which allows a full night of dark for the insecticide to do its work.

Third, *choose efficient spray equipment for your scale*. We started with a small 1-gallon hand pump sprayer that we lugged about by hand. Next we

Crops That We Cover to Exclude Pests

Here are the crops that always receive row cover for at least part of their life for insect control on our farm.

Arugula, baby Asian greens, baby kale. When these crops are grown in the field, we always cover them for the duration of their life to exclude aphids, flea beetles, and other pests. Since these crops are sold as fresh greens, their leaves must be free of holes.

In greenhouses, pest pressure is lower, so we usually forgo covers.

Hakurei turnips. In our conditions, Hakurei turnips must always be covered from seeding to harvest to exclude

We cover turnips from seed to harvest with AG-30 row cover.

We grow cucumbers up trellis netting and cover them for the first few weeks, when they are most vulnerable to pests. After uncovering, we help plants find and attach themselves to the netting.

upgraded to a Solo hand pump backpack sprayer. This sprayer was well built and gave many years of service, though it required considerable effort as our business grew. Now we use a gas-powered Solo mist sprayer, and I would not want to be without it. It operates like a leaf blower with a drip hose on the nozzle, spraying crops with a fine mist. This sprayer is many times faster than hand pump models and uses much less insecticide. These factors alone easily

Zucchini plants covered with row cover to exclude pests. We remove the cover when the first flower appears.

seed corn maggots, which hatch from flies that lay eggs near the base of the plant and then burrow into the roots.

Radishes. We cover all field-grown radishes to exclude flea beetles until harvesttime. In the greenhouses flea beetles are less of a problem.

Kale. We cover kale to exclude cabbageworms, which hatch from butterfly larvae. The adult butterflies, sometimes called cabbage whites or small whites, start fluttering around our farm every May or June. After a while kale plants grow too tall to cover efficiently. At that point we switch to spraying with *Bacillus thuringiensus*, also known as Bt, a bacterium that naturally produces a crystal protein that is toxic to many pest insects (see "Sprays for Pest Control" on page 113).

Cucumber and zucchini. We cover these to keep out cucumber beetles, squash vine borers, and squash bugs, among others. We grow both of these crops as transplants and cover plugs as soon as they go in the ground. We remove the covers a few weeks later when flowers appear, to allow for pollinating, after which point pest pressure lessens and our crops survive on their own.

pay off the investment. The real bonus, however, is performance. It is capable of evenly coating all sides of leaves, including underneath, where many pests lay their eggs. Gas-powered sprayers are heavier, so I suggest trying one out before buying. A number of other brands besides Solo offer mist sprayers.

In practice, because of our success with preventive measures, we now only routinely spray on two occasions: we use a spinosad-based product on

We use a mist sprayer to apply organic insecticide evenly on all sides of leaves.

potato beetles in June and July, and we spray Bt on summer worms. There is not room in this book for a comprehensive discussion of all the insecticide options available for organic farms. When we confront a pest problem that we can't control through exclusion, we refer to the "Insecticides Comparison Chart," available at Johnny's Selected Seeds (link in "Small-Farming Practices" on page 221). In the early days, we would spend hours each week fighting pests, mostly with expensive organic insecticides. Now, with preventive techniques, the sprayer rarely leaves the shed.

CHAPTER 9

Collecting Cash: Leaning Up Sales

Everyone lives by selling something.

—Robert Louis Stevenson

We sometimes track our lean progress by asking a simple question: how many times have we touched our crops? While the number of touches doesn't equate exactly with profits, it does tell us something about the efficiency of our systems. At first, we might have touched a head of lettuce or a tomato 10 or more times by the time we loaded our van. With lean methods, touches are few and lines of work are straight.

The same principle applies to sales. That head of lettuce or tomato should travel from van to customer with just a few motions and with minimal interruption—with smooth flow. Because all markets are unique, there is no set formula for how to create perfect flow when selling. Still, here are three examples to show how we have leaned up sales.

1. Use delivery metrics to choose markets.
2. Set up for smooth flow at the farmers' market.
3. Use "stacked" marketing to eliminate overproduction waste.

1. Use Delivery Metrics to Choose Markets

A lean technique to boost productivity is to define limits, or metrics, to steer your work. Metric setting is a powerful tool to lean up produce delivery.

One metric we use is a target of *$100 of sales per hour* for any type of selling activity, whether dropping off CSA boxes, selling at a farmers' market, or delivering to restaurants. The clock starts when the van leaves the farm and stops when it returns. For example, at a three-hour farmers' market our goal is to sell $300 of product. If a CSA or wholesale delivery

What's a Tomato Worth? The value of a tomato depends on its location. At three hours away from customers, a perfect-looking heirloom tomato might be worth very little. A perfect heirloom tomato delivered to the right doorstep or market, in the right town, might be worth $6.

route takes us two hours, we expect to deliver at least $200 worth of food. Many times we exceed our goal. When we don't, this metric prods us to improve.

Another metric we devised is a *10-mile rule*. We made it a goal to sell as much as we could within 10 miles of the farm. This encouraged us to really get to know local customers, and it cut down dramatically on our drive time. At first, we delivered even small orders hours from our farm. Now, 90 percent of our food is sold within those 10 miles. To get to that point, we had to form close relationships, and we "pushed" a bit, supplying samples to let local customers know what we could produce for them. We were fortunate to find outlets close to home. Many growers will need to drive farther afield. As another example, some growers use a dollar-value-per-vehicle-load metric. This metric says that ideally a vehicle does not leave for a delivery route until it holds $500 or $1,000 or some other dollar-value worth of food. Self-imposed limits like these spur creative thinking. They encourage consolidating delivery routes and becoming more efficient with time, steering a farm toward higher profits.

A Delivery Vehicle for Smooth Transitions

When we started farming we purchased two full-size cargo vans. In retrospect, this was more than we needed. On many trips the vans were barely half full. Big vehicles, we learned, don't by themselves lead to large sales.

A lean delivery vehicle is just big enough.

My advice when choosing a delivery vehicle is to look for large doors, access from all sides, and the ability to turn a tight radius, because many restaurants require alleyway delivery. Of course, any produce delivery vehicle should be air-conditioned. Stay as small as you can while keeping efficient motion—and fuel efficiency—in mind.

We have since downsized to a single Honda Odyssey minivan. We have removed the seats and installed a sheet of vinyl on the floor. A small vehicle suffices because we make several delivery trips per week. The van holds several hundred dollars' worth of produce, yet is small enough to drive anywhere, including up to alleyway doors. It has two side doors that open wide plus a back door for easy unloading.

"Head" Metrics to Steer the Farm

While at the beginning of our farm we filled notebooks with copious records, we have since done away with that work. Instead, we now use simple metrics that we can track in our heads as we harvest, package, and sell food, hence the term *head metrics*. According to lean, any record keeping is a form of *muda*. It does not directly add value. Spend as little time as you can gathering and poring over records.

Here are some of our "head" metrics:

1. *Crops need to yield $2.50 per square foot or better or we don't grow them* (we sometimes grow a small amount of lower-value items for our CSA; see appendix 3).

2. *$35 worth of produce should fit into a 14-gallon tote,* our standard harvest tote. This keeps our focus on lightweight but high-dollar-value items (for a chart, see *The Lean Farm*).

3. *Crops should move from field to cooler at a rate of $100/hour.* That is, a worker can harvest, wash, package, and set into our cooler $100 worth of the crop within one hour.

We occasionally track crop profitability when we change methods or grow a new crop, adding up time and other costs involved in production. There is no need to do this frequently, however, since we repeat our process for each crop.

While these "head" metrics suit us and fit our situation, I recommend choosing your own metrics, based on your own goals. In our case, we wanted a farm where we produced lightweight, high-dollar-value items, and where every square foot was intentionally tended. We designed the metrics to challenge us a bit. We don't expect to meet them every time, but they have become useful goalposts prodding us toward our vision.

2. Set Up for Smooth Flow at the Farmers' Market

We have designed our booth intentionally at the farmers' market to minimize handling our food. First, we rent ample table space—three 8-foot tables—so that setting up is quick. Second, at the end of each market, we stage the booth for the next market, setting out signs and baskets (our market is indoors and our setup is permanent). Third, we sell almost everything in what we call grab-n-go units. We bunch root crops, pre-bag all greens and beans, and box up tomatoes, peas, potatoes, and many other crops into quart containers. This avoids weighing, a time- and motion-grabber. In addition, our customers usually bag their own items. For those who forget

Our goal is to set food on the table and not touch it again. Customers often bag their own food, and we sell in grab-n-go units whenever possible to keep traffic moving.

bags, we set extras at the ends of tables. These systems lessen the need for us to touch our food after it is on the table.

Additionally, we create a professional look that draws people to the booth and tells them we are serious about selling. When we arrive, we set out as much food as possible, and we bring extra lights so that produce is easy to see. Signs are bold and easy to read. We religiously follow the adage "Stack it high and watch it fly," piling carrots, radishes, beets, and turnips in big mounds, and setting out at least a full table of tomatoes in quarts. During the market, we "face" the booth, meaning we push products forward, closer to customers, as spaces open up. These tactics ensure that our food flows quickly and smoothly.

3. Use "Stacked" Marketing to Eliminate Overproduction Waste

Overproduction waste is among the most insidious forms of waste on a farm. If you consistently grow food that never sells, you won't be in business for long. Every unsold pepper and head of garlic saps energy—and money— from your farm without giving back, except perhaps in the form of expensive compost. Larger farms can often mask sloppy production with high volume. Small farms, with tighter margins, don't have that luxury.

No matter how efficiently you grow food, you only receive cash when it sells.

Overproduction waste is among the easiest wastes to track: simply weigh the crops that come back to your farm unsold. In the produce world, this number is called a *shrink rate*. In the first few years of our farm, our rate of shrink was high, as much as 50 to 100 pounds per week. Now we barely have enough leftovers to feed to our flock of six hens.

One practice that accounts for our low shrink rate is lean planning. As discussed earlier, we pre-sell as much as we can, track weekly sales at our farmers' market so we send to market just the right amount, and quickly adjust what we grow to meet changing demand.

Another practice is to develop "stacked" markets: find secondary and tertiary markets for our goods. These are outlets that provide a sort of safety net in the event of overproduction. For example, our largest market of the week is a Saturday farmers' market. While we aim to send in just the right amount, we sometimes have leftovers. In order to find homes for them, I call up wholesale buyers—chefs or our local co-op produce buyer—and offer a Saturday-afternoon or Monday delivery of items still in quality condition. We also restructured our CSA so that most boxes are filled and delivered early in the week. This is another outlet for tomatoes, potatoes, and other Saturday market items that might have gone unsold but that last. This stacked marketing strategy prevents shrink. It ensures that every effort we make on our farm counts. It is a key practice that allows us to earn a comfortable living on less than an acre of land.

CHAPTER 10

Lean Applied to Our Best-Selling Crops: Seven Case Studies

Lean is a way of thinking, not a list of things to do.

—*Shigeo Shingo*

Several years ago a factory owner from North America visited a Toyota factory in Japan. During the tour he took a particular interest in a system for distributing parts, called *flow racks*. He studied the system and took assiduous notes. Wanting to copy the "Toyota Way," he installed flow racks in his own factory as soon as he returned home.

A few years later he revisited the Japanese factory and was confused to find the flow racks mostly gone, replaced by a new system for managing parts called *kitting*. "So, tell me," the flummoxed man asked his hosts, "what is the right approach? Which is better, flow racks or kitting?"

The Toyota workers did not understand the question. They said, "Whenever you visit us, you are simply looking at a solution we developed for a particular situation at a particular moment in time."[1] The answer probably did not sit well with the visitor.

What this story points to is that, at its core, the lean system is a way of thinking, not a particular way of doing. It is about discovering what your particular customers value, and rooting out the particular wastes in your unique value stream. Context is everything. There are no cookie-cutter solutions. That which "flows" on one farm might "flop" on another.

In this chapter I outline specific steps we use to grow our best-selling crops, pulling together techniques from the previous chapters. While my goal is to illuminate basic information you need to know in order to master these crops—for example, tomatoes germinate best between 75°F and 80°F—bear in mind that the case studies are glimpses through a window onto a scene that constantly changes, as our customers change and as we discover new wastes to eliminate. Use the case studies, like the rest of this book, as a starting point for developing your own lean process. For a full list of our plant varieties see appendix 2.

1. Tomatoes
2. Baby greens
3. Kale
4. Head lettuce and romaine
5. Carrots
6. More bunched roots: turnips, radishes, and beets
7. Peppers

1. Tomatoes

In 1521 the Spanish conquistador Hernán Cortés and his army stormed into the Tlatelolco market, on the outskirts of Tenochtitlan, now Mexico City, and brutally upended it. For weeks fighting ensued in what is now called the Siege of Tenochtitlan. Ultimately, the Native Americans, besieged more by smallpox than weapons, could not halt the Spanish advance, and the great Aztec capital fell into European hands.

We sell heritage tomatoes to chefs in 10-pound boxes. Because they bruise easily, we keep them in a single layer. The Marnero variety is our best-selling tomato.

The event was among the most decisive in the conquest of the New World. It opened up giant swaths of territory to the Europeans and gave them substantial control of the Pacific Ocean, enabling them to send loot around the world.

One of those pieces of loot, it so happens, was a tomato, most likely hard, yellow, small, and, to those in the Old World, completely novel. It didn't go over well. While the fruit had been widely revered in Mesoamerica—the Pueblo native people thought eating its seeds imparted powers of divination—the Europeans for the most part thought it stank—literally. The Englishman John Gerard wrote in 1597 that it was a food of "ranke and stinking savour." In the United States the plant breeder Alexander Livingston wrote, "Even the pigs won't eat them."[2] (Livingston later changed his mind and, in the 1800s, became the father of US tomato breeding.) For decades tomatoes were grown only as an ornamental, a boundary accent for peas, potatoes, and peppers.

Times have changed. Tomatoes are now the centerpiece vegetable in most US gardens. My grandmother, and many other grandmothers besides her, grew tomatoes in pots on her porch when she was no longer able to tend a full-size garden. Tomatoes are the best-selling supermarket produce item in the United States. According to the USDA, we consume more than 30 pounds per person per year, a staggering amount considering their initial reception.[3]

On our farm, tomatoes account for more than one-third of our sales. We sell specialty heritage tomatoes to artisan restaurants, red round slicers to local bars for their hamburger plates, yellow tomatoes to our local food co-op, and all shapes, sizes, and colors to our farmers' market customers. They remain the most asked-for item in our CSA boxes. We give them our most focused attention, coziest growing spots, and best compost. Here is how we grow them.

1. *Our tomato growing starts by collecting orders.* Each of our customers, it seems, wants a different type of tomato. Our goal is to grow, every year, precisely what each customer wants, no more and no less. In the winter of 2016, for example, I sat down with the produce manager of our local food co-op to review tomato sales the previous year. Yellows sold almost as well as reds. Of heritage-style tomatoes, dark purple and green were favorite colors. A brewery told me they planned to add a tomato-mozzarella plate to their 2017 summer menu, and they would take 20 pounds a week. They asked me to grow a mix of bright-colored, medium-size tomatoes. These conversations create *pull* on our tomatoes.

   To meet demand for as long a season as possible we start seeds once a month from January to May. Tomatoes occupy most of our greenhouse space in summer. Because quality is always higher in the greenhouse, we no longer grow them outdoors, with the exception of Artisan cherry tomatoes (Artisan Seeds is a company that produces seeds for heritage tomatoes). We grow indeterminate varieties, which grow to an undefined height, and determinate varieties, which stay short. Determinate varieties are easier to grow than indeterminate. Our best-selling indeterminate variety is Marnero, followed by Margold, two heritage tomatoes. Our best-selling determinate varieties are BHN-589, a red tomato, and Carolina Gold, a yellow variety (see the "Heritage, Hybrid, and Heirloom Tomatoes: What's the Difference?" sidebar on page 126). Our best-selling small tomatoes are Artisan Tomatoes.

Heritage, Hybrid, and Heirloom Tomatoes: What's the Difference?

Heirloom tomatoes are open-pollinated tomatoes whose seeds have been passed down for generations. Hybrid tomatoes result when plant breeders intentionally cross two different varieties of a plant. *Heritage* is a term often used to refer to hybrid tomatoes bred for the look and flavor of heirlooms. These tomatoes are newer to the marketplace.

Among tomato fans, heirlooms are usually considered choice. While typically nonuniform, their flavors often surpass hybrid varieties. The downsides are they don't store as well, yield poorly relative to most hybrids, and are prone to disease. Also, heirlooms are indeterminate, requiring more complicated growing methods than determinates.

Marnero tomatoes, a heritage variety. We establish pull on our products by precisely identifying the specific type of tomato each customer wants. These Marneros will be sold to a high-end bar and restaurant.

2. *We aim for no-defect seed starting* with tomatoes, seeding them beginning in late January under the grow light setup in our basement, where we can easily control heat. After March, when we can more affordably heat the greenhouse, we move to germinating them in our propagation house. We use 10-row seedlings flats, spacing seeds 1 inch apart. We lightly cover seeded flats with fine vermiculite and set them on heating mats or into chambers set between 75°F and 80°F. If using heating mats, we top trays with 7-inch-tall clear propagation domes, which can be vented, to help hold in humidity. The more standard 2-inch domes often trap too much heat. We check in on the flats several times per day; the seeds typically germinate in three to five days. Specialty tomato seeds can cost more than $1 per seed—their high cost motivates us to make sure they grow. Once seeds germinate, the domes are removed, lights are placed to within 3 to 4 inches of the seedlings, and we lower nighttime temperatures to 68°F. A week later we drop nighttime temperatures as low as 60°F and keep day temperatures at 65°F to 68°F. This gradual cooling off promotes stockier growth.

Once true leaves form, after about two weeks, we pot them up to 50-cell flats for an additional two weeks, or until roots have filled the cells. Then we pot them up to 4-inch pots. Each time we pot up we bury plant stems down to the first true leaf, sometimes gently twisting them as we do so, to keep them stocky. Tomatoes stems are unusually flexible in this way.

We start tomatoes in 10-row seedling flats under humidity domes (removed for the photo). A pocket thermometer helps us monitor temperature. These tomatoes are under grow lights on a heating mat in our basement. We soon pot them up to 50-cell flats, burying stems.

Tomatoes are best transplanted when the plants are six to seven weeks old. These are in 4-inch pots. With a compost-based potting mix, we do not need to use any other source of fertilizer on tomato seedlings.

Preventing Tomato Disease

Like weeds and pests, plant diseases are better prevented than treated. Diseases common to tomatoes include airborne molds and soilborne viruses.

Crop selection is our most effective weapon to prevent diseases. When I see disease, I first find an accurate diagnosis, consulting our books or texting an image to a Purdue consultant. If I am really stuck, I send a sample in the mail to a plant lab. The next time I order seeds, I look for varieties resistant to the disease. For example, in 2015 several varieties exhibited leaf mold, often abbreviated as LM. When shopping for 2016 seeds, we sought out "LM-resistant" varieties. Careful seed selection costs a fraction of spraying plants for disease.

We also use cultural practices to prevent disease, such as the following:

- Staking and pruning to increase air movement
- Using drip tapes instead of overhead irrigation
- Not working on tomatoes when their leaves are wet
- Using raised beds for better drainage
- Sweeping up aisles of plant debris and dirt
- Regularly washing and sanitizing harvesting equipment

Essentially, we aim to maximize the flow of fresh air and to keep our equipment and spaces clean. While many companies sell sprays and inoculants to control disease, at this point our preference is to rely on variety selection and these cultural practices.

3. *We analyze our moves at transplant time to save effort.* While we used to transplant tomatoes into plastic, now we plant straight into bare soil, in part because it is much faster than wrangling with plastic. Usually two people work together, one pulling plants from pots and placing them in the row, and the other pulling soil apart and popping plants in. Our rows are 5 to 6 feet apart, depending on the greenhouse, and we space plants 9 inches apart in the row for indeterminates, or 18 inches for determinates. We grow them in slightly raised beds with 24 inches across the bed top.

We plant into pre-irrigated soil to save time watering. Immediately after transplanting we run two lines of drip tape on either side of the plants. This waters plants evenly and is a form of insurance: if one tape plugs up, there is a backup. As a general rule, until fruits form we water heavily—between three and six hours per day, depending on conditions. At this stage we use a finger method to gauge moisture. We aim to feel wet but not soppy ground 2 inches below the soil. After fruits form, we water from zero to three hours per day, again depending on conditions.

To Graft or Not to Graft?

Many growers now graft tomatoes to increase vigor and disease resistance. The process involves growing a vigorous hybrid rootstock, bravely slicing off the top with a thin razor, and then grafting onto it a "scion," the top half of your desired tomato variety, using clips or tape. To read more about the process, I recommend Andrew Mefferd's *The Hoophouse and Greenhouse Grower's Handbook*.

While grafting can solve certain problems, the practice comes with a host of costs. To name a few: the time spent grafting, the cost of extra seeds, the expense of a healing chamber and grafting supplies, and the cost of lost plants (even the most experienced grafters lose plants in the process). While we have grafted in the past, we no longer use the practice because most diseases showing on our tomatoes have been airborne leaf diseases; new roots would do little good against them. Second, to achieve higher yields it is faster and simpler, in our case, just to plant a few more tomatoes. Also, new tomato varieties feature ever-increased disease resistance and vigor already built into them. My advice is to consider grafting if you have a persistent, specific disease problem that a rootstock could solve. Otherwise, you may be better off skipping the step.

The back portion of our propagation greenhouse makes an ideal spot for early tomatoes. It is heated to 55°F to 60°F.

As a general rule, when ripe tomatoes burst and crack open—indicating too much water—we dial back our irrigating.

4. *For fertility management, we rely on our own compost,* sometimes amended with minerals, applied monthly as a side dressing. This saves time and money compared with applying sprays and other organic fertilizers, as was our practice in the beginning. Because plants are in bare soil, we have easy access to their bases, where we apply the compost. As stated, tomatoes are the only crop for which we regularly test soil and run petiole samples. If you rely on such tests, take them early. In 2016,

The two-leader method. Plants are set 9 inches apart in a single row with leaders heading in opposite directions.

Indeterminate tomatoes growing diagonally up twine attached to a spool. The spools hang on #9 wire that runs the length of the greenhouse, tightened with ratchet fence tighteners. The wires are 24 inches apart above each bed.

We allow plants to grow to the #9 wire, and then we lower more twine as we push the spools, which are on hooks, down the wire. The technique is called *lean and lower.* Notice we pruned tomato clusters to five fruits. To add compost once a month, we simply roll back the landscape fabric.

samples taken a few weeks after transplant time showed a potassium deficiency. The early test gave us time to add potash to our compost, with quantities guided by a Purdue consultant. Another tip: tomatoes like to take up nutrients in stages, not all at once. As plants get bigger, we give them proportionally more compost, up to ½ gallon per plant, each month, at peak season.

5. *We manage plants for the highest possible yield by pruning and trellising.* Leaf pruning starts when plants are about knee-high, when we clip off the lowest set of leaves to increase airflow. By the time plants are

We use two vine clips per plant to prevent plants from slipping on the twine. Otherwise, we train leaders by twisting them up the twine.

To pollinate, we start off by hand, using this tomato pollinating wand.

For determinate tomatoes, we pound T posts on each end at a 20-degree angle. Then we pound in 72-inch white oak stakes between every two tomato plants and string twine horizontally every 12 inches as plants grow.

After the weather warms, we purchase one Class B hive from Koppert Biological Systems, which services all four greenhouses. Note that it sits on top of a moat to keep out ants.

waist-high we have typically pruned leaves off the entire bottom 12 inches of the plant. We do this for both determinates and indeterminates.

We trellis determinates using a "stake-and-weave" method, with one 5-foot-tall white oak stake pounded into the ground between every two tomato plants. On the ends of rows we use metal T posts driven in at a 20-degree angle away from the rows. This provides a strong post for twine, which we weave between stakes every foot or so as the plant grows.

We grow indeterminates diagonally up poly twine. The twine unravels from spools that hang from #9 wire, tightened with electric-fence wire tighteners, or turnbuckles, strung along the greenhouse from end to end, on top of a 7½-foot crosstie. Two runs of #9 wire hang 24 inches apart above each row of tomatoes, which lie in the middle of the wires. In addition to pruning bottom leaves, we prune suckers that grow between the main stem and side stems, leaving just two leaders, each growing up its own twine, and each heading in an opposite direction, north or south. Eventually plants reach a length of 20 to 30 feet. To achieve a diagonal line, we lower twine as needed from the spools—about once per week during peak season—and push the spools along the #9 wire. Once plants set fruit, the twine sags a bit from the weight, creating a slight curve. When plant tips reach the end of a row, we move their spools over so that the leaders form a loop.

This system decreases leaf disease and increases fruit size, and it also creates a tidy appearance. However, all that trellising and pruning can gobble a lot of time. We save time by forgoing the clips many growers use. Instead, as plants grow we twist their tips around the string. We do use a few clips per plant to ensure they don't slip. Also, we stop all trellising and pruning about one month before our expected final harvest. The plants might appear neglected as the tops flop downward, but by then, because of cooler weather, yields are quickly dropping, and there is no point in spending time tidying just for looks.

To pollinate, we agitate each flower cluster with a tomato pollinator wand twice a week. This releases pollen. In May we purchase a Class B bumblebee hive, designed especially for greenhouse production, from Koppert Biological Systems, and let bees take over this work. We wait until May because by then it is warm enough to open the sidewall curtains on our greenhouses, allowing bees to travel from one greenhouse to another. We purchase a second hive in July to replace the first, which by then is declining. Unfortunately there is no way to perpetuate the hive year-to-year because the life cycle of the bees developed for this pollinating system is too short. More common honeybees can also pollinate tomatoes, but less effectively.

Beware Stems

Stems left on tomatoes easily become little pokers, puncturing other tomatoes and rendering them unsellable. It is best to remove stems at the time of harvest. If you do leave stems on, as required by some varieties to prevent cracking at the top, I recommend harvesting into single-layer boxes to prevent puncture wounds.

We harvest most determinate tomatoes by hand, removing stems as we go, and placing tomatoes one on top of the other into boxes. We harvest indeterminate tomatoes with curved grape shears and leave stems on. Those go single-layer into 10-pound tomato boxes.

6. *At harvesttime, we trim our effort by washing and sorting tomatoes as we pick rather than later.* At first we would pick all of our tomatoes and sort out the mess later in the processing room. This entailed a lot of handling, first to put tomatoes into intermediate containers, and then to pull them out, look them over, and place them in a different box. Now, whenever practical, we carry a damp rag with us and, if needed, wipe tomatoes as they come off the vine, sorting at the point of harvest. All tomatoes then go into their final shipping containers. This adheres to the timesaving principle, "Harvest as market-ready as possible" (see the "Single Piece Flow" sidebar on page 143).

2. Baby Greens

While tomatoes traveled from Mesoamerica to Europe, lettuce traveled in the opposite direction. Lettuce was first cultivated in ancient Egypt so that oil could be extracted from its seeds. It was popular in Europe for centuries—the Romans served it at the end of feasts as a nightcap, since early cultivars contained high levels of turpentine-based alcohols, which make you sleepy. Christopher Columbus, in the late fifteenth century, then brought the first lettuce seeds to the New World. Modern, less sleepy varieties feature sweeter leaves and come in all shapes and colors.

Lettuces are most popular in our markets when harvested at the baby stage and sold as part of a salad mix, also called *mesclun mix*. Other baby greens we sell include arugula, spinach, braising greens (a mix of greens for cooking), and kale. As a group, baby greens represent our next best-selling item after tomatoes.

Baby greens find high demand in many markets. Local growers can provide a crop fresher than trucked-in greens.

Baby greens are an ideal small-farm crop. With greenhouses they can be grown year-round in most places, they yield high per square foot, and skilled local farmers can deliver a product superior to shipped-in greens, which might be a week old or more by the time they reach store shelves. Baby greens fit the lean system well, too. With their fast turnaround, a farmer can quickly scale up or down to match customer demand. Our steps:

1. *We receive orders for greens through weekly accounts whenever possible.* Many restaurants keep greens on a permanent menu, so we establish standing orders when possible. With one account we have delivered 5 pounds of salad mix per week, with the exception of just a few weeks, five years and counting. Even though it is a small order, we are glad to fill it because we can pre-plan, thus cutting overproduction waste.

   To make sure we plant the right amount, we estimate by the bed. For instance, we know from experience that we sell to our farmers' market customers on average one bed per week of baby greens. This doesn't mean we always seed the same amount. In early spring, when growth is slow but demand high, we seed as many as three beds per week. Likewise, going into fall we double- or triple-seed to make sure we have enough stock for late fall and winter. On the other hand, for two weeks on either side of the summer solstice, when plants grow fast and demand dips as other vegetables come available, we cut our seeding in half.

   Our all-around favorite variety of baby lettuce is Defender romaine. It tops all varieties in yield per square foot and holds well in winter greenhouses. Other favorites include Dane, a frilly lettuce that adds loft to the mix, and Red Saladbowl, which performs well in extreme

Baby Lettuce versus Multileaf Head Lettuce

Two basic types of lettuces now find their way into many salad mixes: baby lettuce and multileaf head lettuce. Baby lettuces are lettuce varieties harvested while in the baby stage, before they develop into a full-size head. Because of their widespread use in salad mixes, breeders now develop many varieties specifically for cutting young. Baby lettuces grow quickly—often in just five or six weeks—and can be cut multiple times, hence their nickname "cut-and-come-again" lettuce. We grow them year-round as part of our salad mix.

In spring and fall we also grow multileaf head lettuce. These are full-size heads that feature dozens of uniform, short leaves that cut easily into salad. In our experience, multileaf head lettuces produce leaves with a longer shelf life and more interesting textures than baby lettuces. Their downsides are a slower growth habit and the time it takes to transplant them. We cut transplant time with the paper pot system, and because of slow growth we generally keep them out of midwinter greenhouses, which we reserve for faster crops.

Rachel holding a paper pot flat of 264 multileaf lettuces.

Multigreen 3 and Multigreen 54 lettuces from the Osborne Seed Company grow to a large size and can be sold as full-size heads or cut into salad mix.

Newer multileaf lettuces offer interesting textures and shapes and add loft to a mix.

temperatures. Our favorite multileaf lettuce, a type of head lettuce for salad mix production, is Multigreen 3 from the Osborne Seed Company. It features a dense head with crisp leaves that sells well in our market as an attractive full-size or mini head lettuce, if we don't need it for salad mix. Our favorite braising greens, sold on their own or mixed with lettuce, are red and green mizuna, Tokyo Bekana, Red Russian kale, koji, and tatsoi. Our best-performing spinach is the variety Gazelle.

2. *We accomplish defect-free seeding by direct-seeding crops when they will germinate best.* All of the baby greens we grow are cool-weather crops. We avoid direct-seeding them during the hottest six to seven weeks of the summer. Because their germination is painfully slow in cold soil, we also avoid seeding between October and mid-January.

   While we still usually direct-seed greens with the Jang seeder, we now use the paper pot method to expand their seasons (see chapter 6). By starting seeds in a protected environment for the first few weeks, we can be sure of good germination and early growth during the cold midwinter. Spacing greens properly, usually seven rows per 30-inch bed when direct-seeded or five rows per bed when transplanted, also reduces the chance of defect in the form of mold and mildew.

3. *To further reduce defect, we rake greens every time we harvest.* In our experience, sales from the first cutting of baby greens often just cover costs. Profits start with the second cutting and beyond. Fourth, fifth, and sixth cuttings, if you can achieve them, are almost pure profit. With greens, it pays to tend a small patch well.

   The best way to achieve quality multiple harvests is to rake beds with a tine rake immediately after harvest (see the "How to Cut Greens" sidebar). This pulls out cut leaves that were left behind and it scarifies—lightly scratches—the soil, preventing mold and mildew. The key is to apply the right pressure on the tool—too much will pull up plants, too little and you leave debris behind. We prefer the

We rake greens after every harvest to increase the quality of successive harvests. Photo courtesy of David Johnson Photography.

How to Cut Greens

For the first cutting of baby greens, if the stand is nearly perfect, we use a 15-inch greens harvester made on the farm out of an aluminum frame, polycarbonate sides, and a band-saw blade, a scaled-down version of a commercial greens harvester sold through Johnny's Selected Seeds. It is a simple tool, with no motorized parts, and easy to sanitize.

In most cases, however, on our small scale, we still prefer knives. One reason is that most greens in our salad mix derive from multileaf lettuces, which do not cut as easily with mechanical harvesters. Also, we find that knives, because they allow precise harvesting, work better than other methods on second harvests—cutting regrowth after the first harvest. With knives, we select carefully, leaving bad leaves in the field.

To harvest quickly with a knife, I aim for three or more slices, collecting greens from the three slices in my hand before moving them to a tote. I quickly look the greens over for weeds, bad leaves, and bugs, on the way to the tote. It is faster to sort in the field as I pick rather than later in the wash station. To stay limber, I switch positions about every 10 row-feet, sometimes straddling a bed and bending over, other times crouching and kneeling in from the aisle.

Before every harvest I sharpen our knives with a diamond hone. We stock cut-resistant gloves for new workers to prevent injury until the worker is comfortable harvesting so close to a sharp blade.

Cut height varies by customer. Higher-end restaurants want smaller 2- to 3-inch leaves. Our other outlets are happy with slightly larger leaves. On a quality stand, I can easily hand-harvest 100 pounds of greens in less than 90 minutes.

If you spend more than a few hours each week harvesting greens, I recommend trying out a mechanical greens harvester. One I recommend for small farms is the 15-inch Quick Cut Greens Harvester, sold through Farmer's Friend, LLC. Mechanical harvesters work best on first cuttings. For second and third cuttings, a knife will offer better precision.

For the first cutting we sometimes use this 15-inch greens harvester, which we modified from a model sold by Johnny's Selected Seeds to be smaller and easy to sanitize. Photo courtesy of David Johnson Photography.

Most often we prefer to harvest with sharp knives and quick hands. This allows for more precision and less waste, especially on successive harvests. Photo courtesy of David Johnson Photography.

We sharpen our knives every time we use them. The keys are to apply firm pressure, hold the knife at a 20-degree angle, and use a full sweep so that the entire blade is sharpened. Be sure to swipe both sides an equal number of times. With practice, the task takes just a few seconds.

21-inch tine rake sold through Johnny's Selected Seeds (but an adjustable-width small-tine leaf rake will also do).

4. *When processing, we handle greens as little as we need to.* We generally harvest greens early in the morning, before field heat sets in, into 14-gallon totes, a comfortable size for one person to carry. After harvest, we bring them into our air-conditioned processing area. We rinse them in four-bay stainless steel sinks, using a farm-built net to move them from one bay to the next. Usually two rinses suffice. One option used on many farms is to attach an air hose to the sinks to agitate the water, though we've not found this to be necessary on our scale. Next we funnel greens into a sanitizable tub that is sized to fit inside a washing machine. We bought the tubs at a local farm supply. I have taken the lids off the washing machines and set them on permanent spin cycle. Unless you are comfortable with rewiring electric appliances, I recommend working with an electrician to convert washing machines. We spin for two minutes, using a timer, then shake the tubs and spin for two minutes more. The best machines for this purchase are industrial duty, with higher RPMs. During wet springs we sometimes dry greens further by placing them for a few hours or overnight in bulb crates in our walk-in cooler. Greens are then bagged and sealed with an 18-inch foot-operated bag sealer. We bag most wholesale orders to 3 pounds, and farmers' market and CSA customers receive either ¼ or ½ pound.

3. Kale

Walk about the island of Skye, off the western coast of Scotland, and you will find curious, centuries-old formations made of stones. The curvilinear walls, more than 4 feet tall, with well-defined doorways and gateways, are

Dumping fresh greens into tank, pulling weeds if needed as I dump. We wash greens in stainless steel tanks.

A 1-inch fish net suspended onto a PVC frame allows us to efficiently move greens from one tank to another. We drilled holes into the PVC so that the frame sinks.

Dumping greens into tub.

Transferring greens after spinning.

Sealing with a bag sealer. We try to keep air in the bags as a way to protect the greens. We fabricated a large shelf, attached to the sealer, to support the bags as we seal them.
Photos courtesy of David Johnson Photography.

too long and irregular to be building foundations, yet not long enough to be fences. In fact, they are ancient "kailyards," protected gardens devised specifically for growing kale, the island's most popular vegetable for hundreds of years. Actually, kale was the most widely eaten vegetable in all of Europe before the higher-class head cabbage appeared sometime in the Middle Ages. The botanical name for kale (*Brassica oleracea* var. *acephala*) literally translates to "cabbage of the vegetable garden, without a head." In Scotland, the word *kail* doubled as a word for "dinner," and each house had a *kail* pot for boiling down the greens before every meal.

After several hundred years in near obscurity, kale, a nutritional powerhouse, is back. We sell nearly as many bunches of it as we do carrots. Part of its popularity lies in its versatility. Kale pulverizes well in juice, cuts up neatly into salads, and cooks up quickly as a braising or boiled green. It can even be baked and turned into chips, tossed with a bit of Parmesan cheese and sea salt.

As mentioned, baby kale is a part of our salad mix. We also grow full-size plants. Our best-selling varieties are Black Magic (a lacinato type) and Winterbor. We seed our first kales of the year during the last week of January, one seed per cell in 72-cell trays, for March 1 transplanting into unheated greenhouses. The seedlings grow alongside our early tomato seedlings. Like tomatoes, they prefer germinating temperatures of 75°F to 80°F, followed by cooler conditions, under 68°F, after emergence. One month later we seed another crop that we transplant April 1 in the field. For winter harvest we transplant the first week of September into both heated and unheated greenhouses. Here are four ways we trim waste.

1. *To reduce defect waste we give kale ample space*—two rows per 30-inch bed, with 9 inches between plants—because our markets prefer large leaves. We transplant all kales into landscape fabric or plastic, adhering to our two-month rule, as kale can occupy ground for up to nine months. We have experimented with both white and black plastic and have not seen a difference in overall yield, in part because kale leaves soon shade the plastic, mitigating its effects.

2. *To achieve large plants and dark-colored leaves, we apply a thin layer of brown high-N compost at transplant time.* We have on occasion reapplied compost to aging kale to give it a second wind. The nitrogen-rich compost ensures deep green foliage.

3. *To prevent pests, we cover kale for as long as practical with row cover.* Eventually kale grows too tall to effectively cover, at which point we switch to spraying as needed with Bt to kill worms.

   The trick in our area is to tend kale well so that it survives the summer, yielding a free fall harvest from spring-planted kale. One way to

Kale planted two rows per 30-inch bed. Note the row cover in the aisles. We cover for both frost protection and insect exclusion, and uncover for harvest, until the plants grow too tall to efficiently cover.

keep kale healthy, besides supplying it with compost, is to keep it consistently watered through drip tapes. Another is to strip off and rake up diseased leaves every few months.

4. *For efficient harvesting, we pull kale leaves by hand and bunch with rubber bands as we harvest.* We place bunches into their final containers in the field, using the single piece flow principle (see the "Single Piece Flow" sidebar on page 143). We grow Swiss chard, though a much smaller amount, using the same process.

4. Head Lettuce and Romaine

Head lettuce and romaine are easy crops to grow, but with a relatively low price point in our area they require a lean approach in order to garner a profit. We include head lettuce and romaine in CSA boxes and sell a steady amount at our farmers' market and to wholesale accounts. We seed every week, using our 68°F germination chamber, to maintain a steady supply. We trim waste in three ways.

1. *We trim defect waste through variety selection.* Every year breeders improve traits in head lettuce and romaine. We typically grow four or more new varieties every year to find hardier varieties popular with

customers. We try to offer a red and a green leaf lettuce most of the year. If heads are smaller we package two or three together. Mixed colors make an attractive display at market.

A challenge with lettuce and romaine is quick bolting in warm weather. We use Coastal Star as a bolt-resistant romaine option. With both lettuce and romaine, however, our favorite option is to avoid the problem altogether by planning ahead to avoid harvesting in July and August. There is plenty of other work to do that time of year.

2. *We trimmed waste in the field by switching to the paper pot system.* We grow all head lettuce and romaine from pelleted seed, three or four rows per bed, depending on size of head desired, using 6-inch paper chains. For jumbo heads we fill every other cell to achieve 12-inch spacing. If you use plug flats, I recommend flats with 128 cells.

3. *We also found ways to minimize moves with field management.* Spaced three or four rows per bed, head lettuce and romaine are quick to cultivate with the moon hoe. Instead of using drip tape, we rely on overhead sprinklers, which are faster to set up. Both lettuces and romaine require steady water. On rare occasions we see slugs, but not often enough to warrant action beyond pulling them off by hand at harvesttime. We harvest with serrated knives, cutting just below the lowest leaf for

Romaine lettuce and head lettuce (*foreground*) alongside other spring crops.

Single Piece Flow

In lean factories, managers use the principle "single piece flow" to eliminate motion waste. This means designing systems that allow workers to add as much value as possible to the item in their hands before setting it down. For example, in a single piece flow setting, when a worker is preparing steel for assembly, the person might cut, de-burr, weld, and label a piece of steel before setting it down. In a more traditional factory, a different person might perform each step, involving a lot of picking up and setting down—a lot of batches. (For more on the differences between single piece flow and batch and queue production, see my earlier book, *The Lean Farm*.)

We use single piece flow when harvesting head lettuce, kale, carrots, turnips, and many other vegetables. Whenever possible, we harvest, sort, pick clean, band—and whatever else is needed to prepare a crop to sell—in the field, leaving messes in our plots, not in our processing room. We call this "harvesting as market-ready as possible." Granted, sometimes it is too hot for field processing—we just need to get crops inside. But whenever possible our approach is to process crops in the field. If you are used to batch-style harvesting, time yourself using both methods, and choose whichever is faster. On some farms the difference will be greater than others, depending on the setup. When we timed ourselves, we were surprised at the results.

We use the principle, "Harvest as market-ready as possible." This means sorting, cleaning, banding, and doing whatever else we can in the field to prepare a crop for market. This leaves messes in the field. Here I sort beets as I pick. Photo courtesy of David Johnson Photography.

Here a worker harvests lettuce, pulls off bad leaves, cleans the base in water, and bands or bags for market—all in one sequence.

full-size heads. We pull off bad leaves, dunk the roots in water, and tie size 64 rubber bands around them to help them hold their shape.

We use the same production methods with bok choy and kohlrabi, as they have similar growth habits and spacing needs. One difference is that we cover them with row covers until harvest to prevent pests. These crops also require warmer temperatures, 75°F to 80°F, for optimum germination.

5. Carrots

Carrots are among the most challenging crops to grow, but they bring high rewards. Who can resist a freshly dug, sugary sweet carrot in December? Demand for carrots in our markets is year-round and consistent. All of our systems must be working well if we are to grow them well. Here is how we've leaned up carrot production.

1. *We avoid defect waste by placing small seed orders throughout the year*, close to seeding time. We have found that carrot seeds perish relatively quickly, especially in pellets. Spending a few extra dollars for shipping pays off. We most often use pelleted seeds to enable precise spacing.

   With the high demand for carrots, we aim to supply them year-round, selling fresh carrots from June to November, and relying on storage carrots from December until May. Our favorite early-season and summer carrot is Nelson, followed by Yaya. Our fall and winter carrot is Bolero.

2. *We seed all year round* in order to offer a consistent supply for our customers. We begin with seeding in late February or March, as soon as the soil can be worked. This early seeding avoids weed pressure from summer weeds. We continue seeding every three or four weeks throughout the year, with our final seeding of Nelson or Yaya carrots occurring in August, for fall harvest. Because of the time they take to mature, we no longer grow carrots in greenhouses, with the exception of one bed seeded in December or January for extra-early sales. Carrots germinate in a wide range of soil temperatures, as long as the soil stays wet until germination. In our conditions, the Jang seeder, with roller LJ-24, spaces carrots about 1 inch apart. If we are worried about poor germination, such as during a windy or dry spell, we double-seed, rolling the seeder over the same area twice, to dispense twice as many seeds.

3. *We grow carrots in loose, well-blended soil.* To aerate, we chisel-plow or broadfork carrot beds as deep as we can. Then we blend the beds deep with our tractor and rotovator or BCS tiller. This loosens the ground,

making it easier for roots to grow straight. Next, we shape the beds as deep as we can with our bed shaper, sometimes pulling additional soil from the aisles by hand, further loosening the soil for straight roots.

Our ground usually holds enough nutrients that we don't need to add compost for carrots. However, we have found that if carrots follow a nitrogen-gobbling crop such as kale or spinach they benefit from a 1-inch layer of compost blended into the bed.

Lifting carrots with a root digger. If the soil is dry, we water the night before we harvest to help the implement loosen the carrots.

A root digger for lifting carrots, made by a local machinist to fit our tractor.

We sell most carrots in bunches with tops on to showcase their freshness. We tie them with produce ties that read "Locally Grown."

We harvest storage carrots in a tilt-bed utility vehicle and dump them for hosing onto a propagation tabletop turned upside down. This technique is an efficient small-scale alternative to a root barrel washer.

4. *We keep the soil surface moist until seeds germinate.* With carrots, there is no exception to this rule. Strategies vary by season. In spring, when the ground is already wet and rains are likely, we water seeds well and then cover with multiple layers of row cover. This holds moisture in and warms soil for two or three weeks until the seeds pop. After this, we grow the carrots for the next four or five weeks under a single layer of AG-30 row cover, which allows in more light. We use timers with overhead irrigation to ensure consistent moisture until harvest. If needed, we remove the covers and water straight onto the beds.

5. *We cultivate carrots early and often.* Typically, our first weeding occurs at the white thread stage, when weeds are just visible, using wire weeders. In a few weeks we return to the bed with wire weeders, moon hoes, and our hands. After about four to six weeks, our carrots can usually

fend for themselves. In addition to hoeing, we use the weed prevention strategies outlined in chapter 8, paying particular attention to the first strategy: direct-seed into your most weed-free ground.

6. *To harvest efficiently, we use our Kubota with the root-digging implement.* We start to dig about a foot in front of each bed so that the digger is well submerged before reaching the carrots. We irrigate carrots for a few hours the night before if the ground is hard so that they release and wash clean with less effort. To fill small orders of fewer than 50 bunches we dig by hand with digging forks.

We hose carrots soon after we dig them, before soil dries on the roots. We sell most carrots in bunches with the tops on to showcase their freshness. To store carrots for fall and winter harvest, we remove tops in the field and harvest into our John Deere Gator. We hose them off as we dump them onto a propagation tabletop turned upside down, where we sort and then hose them again. Larger farms use mechanical root washers, such as root barrel washers, to wash root crops, but the investment would not pay off on our scale. We store carrots for winter in our walk-in cooler, in open containers that we keep covered with wet towels. As long as the towels stay wet, the carrots stay crisp.

6. More Bunched Roots: Turnips, Radishes, and Beets

Fresh, colorful root crops, displayed with roots facing up, always help our farmers' market booth to pop. Turnips and radishes are also popular with our CSA customers as grab-n-go snacks that can be tucked into their children's lunch boxes with no prep. I group these crops together because we grow them with similar methods, and we sell them in the same way—banded together in bunches, sold with the greens to show off their freshness, for as long as the greens look nice.

While profitable, these crops require a lot of motions to grow. These lean practices help make them worth their while.

Turnips. As stated, we grow turnips with the paper pot method, in clusters of three or four seeds per cell. The only variety we grow is Hakurei, as it grows quickly and tastes sweet when eaten fresh.

We wasted a lot of effort early on by pushing the envelope too far: planted too early, turnips won't grow fast enough to form nice roots; planted too close to summer, and their roots become woody and bitter. Most years, the target windows in our area for planting outdoors are

Beets grown 2 inches apart using the paper pot method.

late March to mid-May, and early September. We also seed a round of early turnips for the heated greenhouse on the first week of February.

For field preparation, we make sure the ground is loose but we don't bother to raise beds as high as for carrots. We give turnips both brown and black compost when we transplant them. Without exception turnips must be covered from seed to harvest with row cover to prevent maggots from boring into the roots. Turnips need a steady supply of water. With overhead irrigation and timers, we make sure the soil stays moist, and use our fingers to test moisture levels. Ideally we feel moisture 2 inches below the soil surface.

Some customers prefer big roots and others want a more medium size, so we bunch them by size as we pick. We pull off rotten leaves and then band them in the field with ½-inch-wide produce twist ties before hosing them off. How big is a bunch? That's up to customers, and we also follow standards others are using in our market. In general, our bunches weigh about 1 pound.

Radishes. We primarily grow two types of radish, the ever-popular D'Avignon and Rover, a standard round radish. As with turnips, we take a conservative approach when deciding when to direct-seed. Since radishes require cool weather, there is no benefit to seeding them too close to summer. They also require at least 45°F soil temperatures to germinate, so there is no point direct-seeding them too early in the year. Our windows to direct-seed are usually March to May and the first two weeks of September.

Fresh golden beets.

When field conditions are optimal, we direct-seed with the Jang seeder at five rows per bed. Radish seed size varies considerably, so I recommend experimenting with different plate sizes with new varieties. Otherwise we use the paper pot method, as described in chapter 6.

Since radishes are a light feeder, we don't add compost. Because they grow quickly, they make an ideal "understory" crop beside potatoes and peas. They can be harvested by the time the main crop needs more room. Like turnips, they require row cover from seed to harvest to prevent maggots.

Beets. Beets require slightly warmer soils than turnips and radishes. We usually start seeding in April for outdoor plots, or a month earlier for an early harvest from a heated or unheated greenhouse. We grow Red Ace and Touchtone Gold.

We transplant beets using the paper pot method, or for small plantings, we grow them in 72-cell plug flats, four to six seeds per cell, and transplant by hand. Since a beet seed is really a cluster of seeds fused together, it is impossible to predict how many seedlings will develop. Fortunately, few pests bother beets, and we have never had to cover them except to avoid a frost. Our beets grow best with a mix of high-N and low-N compost slightly incorporated at transplant time. As with turnips, we aim for consistent soil moisture. We sell beets with tops on unless they are blemished, in which case we cut the tops and sell the beets in pints or quarts. In that case, we leave ½ inch of the top when we cut to prevent "bleeding" from the wound.

7. Peppers

Peppers, like tomatoes, require several months from seed to harvest and require a similar amount of room per plant, with nearly the same growing conditions. However, they yield less than tomatoes, making them less profitable per square foot. They really shine at harvesting time, though. They are faster to pick than almost any other crop, and if grown in a greenhouse, they usually require no cleaning. We sell almost all of our peppers as colored peppers. Colored peppers sell for three times as much as green peppers, a justified return for patience.

Our standby varieties are Red Knight, Early Sunsation, and Purple Beauty, providing a colorful red-yellow-purple mix. We seed them the last week of January, alongside our first tomatoes, for transplanting into 55–60°F greenhouses in early or mid-March. We germinate them at 85°F, and after germination we grow them with our tomatoes, between 60°F and 68°F. We pot them up from 10-row seedling flats to 50-cell plug flats about two weeks after germination. A few weeks later, we pot them up again to 3- or 4-inch pots. We are careful not to bury stems, because unlike tomatoes, they do not grow adventitious roots through their stems. Also, we keep newly potted-up peppers in the shade for a day, as peppers don't take transplanting well. We transplant them to the field when they are between eight and nine weeks old, two rows per 30-inch bed, with 24 inches between rows and 18 inches between plants. Here are lean tips.

1. *We trellis with a stake-and-weave system.* We use a modified stake-and-weave system to trellis peppers, placing a 5-foot oak stake between every three plants. As with tomatoes, we use metal T posts on the ends for stability. Once plants are knee-high, we weave twine on both sides of each plant to hold them upright. If needed, we add a vine clip (the same kind used for trellising tomatoes) to further stabilize plants. As the plants grow we add more twine but eventually only on the outside—this creates a box, and the plants support themselves on the inside of the box. This configuration allows us to pack a generous number of plants into a small space.

   We used to train peppers up strings in the greenhouses, pruning to two leaders per string, much like tomatoes. While this is a common practice, we found it required too much effort compared with the stake-and-weave method for our relatively short season. If we were willing to heat for a longer period of time, perhaps for nine months or more, trellising in this manner might pay off.

We stake peppers similarly to determinate tomatoes. Note shade cloth to prevent direct sun from scalding ripe peppers.

We sell most peppers as a mix. Colored peppers sell for a higher price than green peppers, and justify the patience needed to wait for them to attain color.

Peppers are relatively light feeders. We give them a thin layer of compost at the beginning of the year and usually no other fertility later on. They share the same drip tape main line and watering schedule as tomatoes.

We prune off the first fruit to encourage the plant to mature further before pushing out more peppers. Beyond that, we do no pruning unless plants become overly bushy.

2. *We use beneficial insects to prevent pests.* Every spring we can count on aphids crawling up and down our greenhouse pepper plants. To control them, a few weeks before transplanting time I call a vendor of beneficial insects and tell them when and how many peppers I plan to transplant. They recommend a species and amount, and we set up shipping dates to correspond to my transplant dates. I have successfully used ladybugs, lacewings, and a small parasitic wasp called *Aphidius colemani*. All three feasted on the aphids, and outperformed organic sprays.

3. *We use a shade cloth to prevent defect from sun scalding.* The primary disease problem we experience with peppers is scarring on the surface caused by too much direct sun. This only happens to peppers planted along the outer edges of the greenhouse—in the middle of the greenhouse other plants shade peppers. On these outer rows we attach a shade cloth with zip ties to the greenhouse where the curtain is rolled up. This extra step keeps the defect waste at bay, increasing our yield on outer rows by about 25 percent.

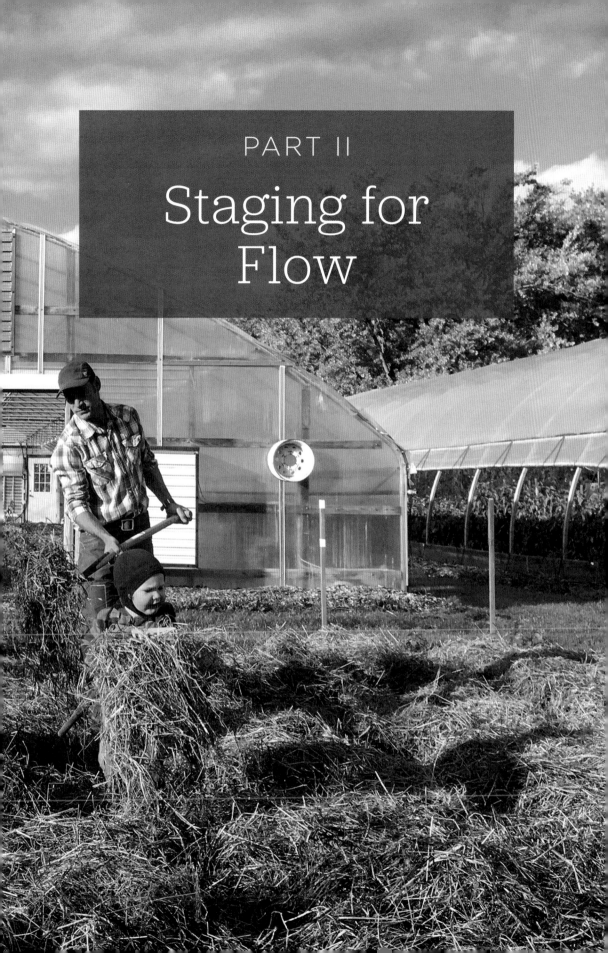

PART II
Staging for Flow

CHAPTER 11

Finding Good Land

In the spring, at the end of the day, you should smell like dirt.

—MARGARET ATWOOD

So far in this book, I've primarily focused on process improvements—doing more with less by trimming out waste. Behind every process, however, is a stage. Just as the smooth performance of a play requires a set, lights, and a building, so smooth flow on a farm requires land, the right infrastructure, and an efficient layout. This is the focus of part II.

The soil on the farm where I grew up—Howe, Indiana—was sandy with a bit of loam. This was the first land I used to grow vegetables, for the small CSA I ran in college. I never worried about standing water. The ground was easy to work because it never clumped together. Carrots came out straight as a pencil. I experienced a few challenges with the soil, too. I went on two trips that summer. Each time, I returned home to a desert; the garden just could not hold water. The location of the garden was also a hurdle. Two sides were open to hundreds of acres of bare land, and wind whipped my plants about, more than once whiplashing crops until they perished.

The next plot I farmed was sandier still, on the communal farm where Rachel and I lived with friends. The farm sat in a crook of the Elkhart River, and the land was almost pure sand. Because it drained fast, we were always able to put in peas, spinach, carrots, and lettuce by March when other farmers were still waiting for their own land to dry out. In spring and fall, when rains fell consistently, our crops flourished. Yet summer crops such as zucchini, tomatoes, and peppers often withered, even with the use of thick mulch to hold moisture in the ground.

What's more, we had no electricity, hence no refrigerator. We hunted for alternatives. For instance, we tried to cool spinach by stuffing it into plastic containers that we tried to sink in the river with cinder blocks. We failed. In the end, we either watched crops wither or drove them into town to a rented cooler.

Every plot of land, it turns out, teaches a lesson. The main lesson we have learned is that good land contributes value while poor land contributes *muda* every single day. If you have the luxury of choosing the land on which you farm, these are the features to look for:

1. Proximity to markets
2. Well-drained soil
3. Generous sun exposure
4. Relatively flat ground
5. The right size for full use
6. Access to water and electricity

1. Proximity to Markets

Direct-market farming means connecting every week with customers, in person. While the idea of farming a plot far away from town might seem romantic, it can cripple your efforts. We drive food to markets at least three times each week. Road time, while usually necessary, is time away from growing food.

The cost of road time is easy to calculate. Take your hours on the road, multiply by an hourly wage, and then assign a mileage cost, say 50 cents a mile to cover gas and maintenance costs. We often use hired help to deliver food so that we can remain on the farm. If the person makes three trips per week, that adds up to about four road hours, counting the time it takes to shuttle around the city. The cost of those trips, counting both mileage and labor, amounts to about $90 per week, if we pay the driver $15 per hour, considered a good wage for drivers in our area. If our farm were located four times as far from markets, our weekly costs would be $360, a difference of $270 per week. Assuming 50 weeks of delivery each year, that totals $13,500 in added costs. In addition, if we were doing the driving, we would be robbed of an additional 12 hours per week. That is 520 days—almost a year and a half—over a 20-year period, during which we could have been growing food or taking a break.

2. Well-Drained Soil

Drainage, the rate at which water physically moves through soil, is key for three reasons. First, roots need air and oxygen, which they don't receive

Your choice of land is the most consequential single decision you will ever make on your farm. Proceed slowly.

Web Soil Survey to Learn about Land

The USDA's web soil survey is the largest natural resource information source in the world, and access is free (link in "Small-Farming Practices" on page 222). We have analyzed reports for every plot that we have farmed, taking note of where one soil type ends and another begins, water tables, slopes, suitability for building, and more. Your local extension agent can help you create, print, and analyze reports relevant for your particular needs, or you can access and print the information on your own.

when sitting in saturated ground. Second, wet soils, especially those with clay in them, stick to equipment and shoes, making farmwork difficult. Third, ground that won't drain will shorten your season. We have farmed plots that did not dry out until late June and that plugged up again in September.

At the other end of the spectrum are extremely sandy soils that drain excessively, denying plants water and nutrients. Though you can bulk up these soils with compost and cover crops, you will likely find yourself constantly adding nutrients to keep crops healthy. One reason we like our clay-loam ground is that nutrients stick around. Ideally, your plot avoids the extremes—pure sand or pure clay—and contains a blend of sand, clay, and perhaps silt or peat.

The quickest way to assess drainage is with your eyes. Do puddles stagnate for days after a rain? When you dig a shallow hole, does it fill with groundwater that won't go away? Find out when other observers have seen standing water on the land. A fine resource is the USDA's Web Soil Survey (see the "Web Soil Survey to Learn about Land" sidebar on page 157), which gives a drainage rating to all land in the United States.

If you do encounter ponding, the land might still work. Occasional ponding is normal. During a downpour even well-drained sands will pond in divots and swales. Greenhouses, tarps, and raised beds all combat excess rain, and we use all three methods on our farm. The problem is when ponding persists for days on end.

Ponding might also result when the water table, the surface level of water in the soil, rises. The USDA's Web Soil Survey indicates expected water table levels. The only way to combat high water tables—which can occur in any type of soil, including sand—is through drainage, typically through perforated plastic drain tiles installed 2 to 4 feet underground. On our current farm, while we did not see persistent ponding in our growing area, we did want to speed up drainage. With a trencher I buried 4-inch plastic tiles every 14 feet underneath our 1-acre growing area. For commodity crops like corn and soybeans tiles are typically spaced much farther apart. Tiling with mechanical trenching equipment is not difficult if you are comfortable with equipment and if your land slopes in one direction. However, this job is often hired out when slopes vary.

3. Generous Sun Exposure

The sun is the heart of your operation. It is the key energy source you are converting into food. You want as much sunshine on your plants as possible. Shadow lines of trees and buildings vary according to season and time of day, but in general an object will cast a shadow twice the length of its height. I recommend placing your garden at least as far away from trees and buildings as their heights. In other words, locate crops 20 feet away from a 20-foot-tall building, and farther if you can.

Which Is Best: Sand, Silt, Clay, or Peat?

In general, sandy loam is the easiest soil type to begin with. It drains fast yet holds on to more nutrients than pure sand. It does not easily compact. Also, assuming it is relatively rock-free, it is a pleasure to work. However, almost all soils, except those with low spots or in contaminated areas, can be amended to grow great food. Some will just take more effort than others.

Table 11.1 shows the major soil types in North America, their pros and cons from the point of view of a vegetable grower, and options for amending them.

Table 11.1. Major Soil Types: Pros and Cons

Soil Type	Pros	Cons	Amending Options
Sandy	Drains fast, easy to work (doesn't stick), loose for good root growth. Warms quickly in the spring.	Requires lots of irrigating and constant fertilizing. Cools quickly.	Cover crops and compost to build organic matter, which holds water and delivers nutrients.
Sandy loam, a mix of sand, silt, and clay, in that order	Drains fast, loose for good root growth, holds nutrients better than sand. The most desired native soil for most vegetables.	Usually still requires fertilizing and organic matter.	Cover crops and compost to help nutrient and water retention.
Silty	Tends to be fertile. Retains water (like clay).	Compacts easily and can become waterlogged. Does not hold nutrients as well as clay.	Organic matter to improve aeration. Avoid compaction.
Clay-loam, a mix of clay, sand, and silt	Drains slowly but usually adequately. Holds water and nutrients well.	Can become waterlogged.	Organic matter to improve workability and drainage.
Clay	Holds water and nutrients better than other soil types.	Poorly drained and can become waterlogged. Compacts easily, and sticks to equipment.	Organic matter to improve workability. Tiles to improve drainage.
Peaty or mucky	High fertility. Holds water well.	Soils usually found in lowland areas with high water tables.	Tiles or canals to improve drainage.

4. Relatively Flat Ground

There are three reasons to farm on relatively flat ground, if such land exists in your area. First, machinery is much safer to operate. Rollover injuries, where farmers are pinned under machines, are sadly common. Even in Indiana, a mostly flat state, tractor rollover accidents accounted for 28 fatalities in 2015, 39 percent of all farm fatalities in the state.[1] With intensive vegetable growing you might traverse a small plot with machines dozens of times per year, a lot of exposure to risk.

Second, heavily sloped ground is prone to topsoil runoff during rains, even when there are crops in the ground.

Third, from a lean point of view, sloped ground will require you to work harder every day. Lean practitioners rate the *muri*, or burden, that workers endure in day-to-day work. For example, an engineer might rate workstations in an automobile assembly line, measuring the weight workers lift and the angle at which they are lifting. If you farm a hillside, trudging up and down every day, your work is packed with extra effort, and your body must compensate to bear that burden.

An ideal plot is actually not completely flat but gently sloped a few degrees south. South-sloping land heats up faster than other slopes because it receives more direct sun for a longer part of the day. This gives you an edge in the early spring. One of the selling points that convinced us to buy our current farm was its gentle slope to the south. However, my experience is that this is really a minor point. Many farmers grow excellent crops on gentle slopes facing east, west, and north.

5. The Right Size for Full Use

According to lean thinking, land, buildings, and machines should be fully used. These are fixed costs. You pay for them no matter how much money your farm produces.

For sure, many growers might value woodlands or idle spaces for their intrinsic value. If you want to preserve a piece of earth and can afford to do so, then by all means do so. We keep a small area where we don't meddle, and let nature go wild. But when purchasing or renting land, I suggest you research the costs, primarily taxes and maintenance, of unproductive land before deciding to buy.

Of course, not having enough land can pose problems, too, leaving you to scramble every spring to find room for your crops. However, a more common mistake I see on small-scale vegetable farms is farmers starting out strapped paying for more land—and buildings and equipment—than they really need.

A quarter acre is a good place to begin for one person growing intensively cropped vegetables. Only if your vision is to get big quickly, with a large crew of workers, or if you plan to grow a lot of sprawling crops like watermelons, might you need more than a few acres in the beginning stages of a market farm.

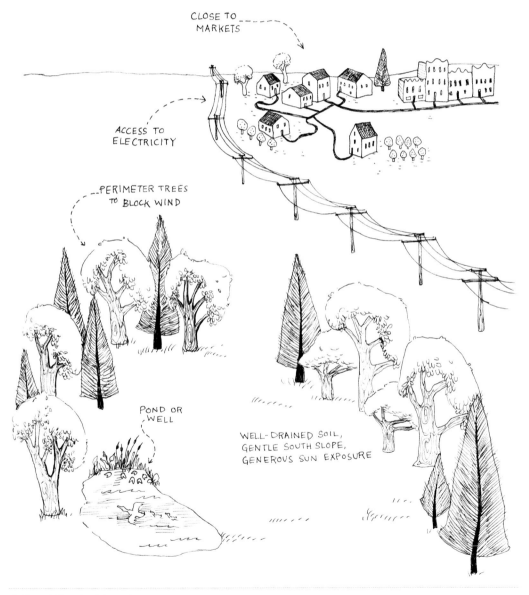

A Good Spot of Land

Four More Factors in a Land Search

Here are more factors I suggest considering when searching for land.

Natural windbreaks. In northern Indiana, fields were commonly surveyed to a size of 20 or 40 acres, separated by rows of trees, called *windrows*, left standing between them. Windrows served a purpose. Indiana farmers before the latter part of the twentieth century grew dozens of types of vegetables and fruits, not just corn and soybeans, and those crops needed protection. Now, sadly, throughout the Midwest thousands of contiguous acres sprawl without a single tree in sight.

On our farm, buildings and trees slow winds from the north, east, and west. The south side is open to neighboring fields, and wind velocities can surpass 60 miles an hour. On that side we have installed snow fencing to buffer against wind. Another option would be to plant fast-growing trees as a barrier.

Rock-free soil. Overly rocky ground wears down rotovators and other equipment and causes root crops to grow into odd shapes. While you can extract rocks with a rock puller implement behind an ATV or tractor, unfortunately this is not a onetime job. Those on rocky ground swear that rocks reproduce and grow like weeds. Avoid them in the first place, if at all possible.

Buildings on the site. While existing buildings can save you time and money compared with putting up your own, inspect them carefully before you buy. Do they really serve your needs, or will they be a burden requiring yearly maintenance? I recommend hiring a building inspector. Inspections usually cost less than a few hundred dollars and can reveal problems you did not see, such as insect infestations or weak foundations. You can also use the inspection report to negotiate a lower price. We have hired inspectors on a couple of occasions and I am always surprised at how many problems their reports uncovered that I did not notice on my own walk-through.

Contamination. Although a piece of ground might look pristine, it might harbor hidden contaminants. In cities, soils are often poisoned by lead dust and other trash from demolition projects. When factories move, they often leave pollutants behind them, creating "brownfields" unsuitable for growing.

Also beware of possible pollutants from neighboring farms. Many farmers in our area use Roundup Ready crops sprayed from airplanes flying overhead. Even a slight breeze causes drift. I know of a gardener in Iowa whose entire garden was wiped out during peak season from drifting herbicide. Also, if your plot is downhill from a field where pesticides and herbicides are used, those chemicals will leach onto your land during heavy rains. To be safe, install a wide grassy barrier between your fields.

6. Access to Water and Electricity

A final component of productive land is access to water and electricity. Both are often overlooked during a land search, but both are critical to market vegetable growing. For water, you can either pump from a pond or river, as long as they are close to your plots, or drill a well. I suggest calling a well-drilling company that services the area around your farm to ask about the potential of a well. Rainwater systems are an excellent backup, but few growers will find them adequate as a main water supply. This step should precede buying your land. I know of a case where a grower bought land and later hired a well driller several times to try to drill a well, to no avail.

Electricity is required not just for the well but for the processing room cooler, scales, lights, heat mats, and myriad other devices on the small farm. Some growers have installed enough solar panels for their electrical needs; even so, many solar systems require a tie to the electrical grid. If you are unsure whether power can be supplied, call the power company and ask to meet an engineer at your property.

There is no perfect piece of property. You might find land that drains poorly but is located close to eager markets. Or you might find loamy soil and a set of buildings but steep slopes and no wind barriers. Should you buy or pass?

All property searches are exercises in compromise. As a final piece of advice I suggest making a list of your top eight priorities. As you tour prospective farms, give a numerical rating to each, on a scale of 1 to 8. Tally the numbers and see which farm comes out on top. Real estate purchases should involve both head and heart; with this discipline you won't forget your head.

After years of renting, Rachel and I took our experience with us in 2008 as we searched for our farm. We toured several properties and settled on a 5-acre former dairy farm, 6 miles from town. The farm was farther from markets than our ideal, and the land was slow draining. But it also featured a well, house, and barn, and the sunny gentle south slope. We had found our blank canvas. It was time to fill it.

CHAPTER 12

Infrastructure and Farm Layout

Organize, don't agonize.

—NANCY PELOSI

It wasn't long after we had moved to our farm that we realized there was work to be done before we could start growing. Posts in the barn were leaning over, in need of support. Paint was completely peeled from the old chicken house, leaving exposed bare wood. There was a well but no water hydrants. And while the barn could house animals, it was not set up for processing vegetables. Thus we decided to spend that first year on infrastructure: building greenhouses, installing drain tiles, remodeling a processing room, building a driveway, and otherwise organizing the farm. We set ourselves up for smooth flow for many years ahead.

Eight Pieces of Infrastructure

Based on our experience, here are eight pieces of infrastructure you need to establish, followed by lean tips for laying out a farm.

1. Processing room
2. Spray station
3. Cold-storage rooms
4. Storage shed
5. Gas house
6. Red-tag room
7. Irrigation system
8. Vehicle access lanes

1. PROCESSING ROOM

A small-scale vegetable farm requires a processing room about the size of a one-stall garage. It should be well lit and easy to sanitize. Ideally, the walls will be covered with panels that can be hosed and the floor finished and easy to mop. It should be insulated and heated if you plan to grow all winter.

The processing room on our farm is a former milking area. When we bought the property we removed 11 stanchions and filled in the manure trough. Later we insulated and painted the walls and added a small gas heater. We use our processing room to wash greens in tubs and pack orders into bags or boxes going to customers.

Inside the room is a handwashing sink with soap. We ask workers to wash hands before handling food at this stage. We use a small electric hot water heater to slightly warm wash water in the winter, and a four-bay stainless steel sink, with bays measuring 20 by 28 inches by 18 inches deep. We use the sink to wash baby greens and microgreens. As an alternative to sinks, some growers use sturdy stock tanks and even porcelain-enameled bathtubs. Above our sink is a 1-inch water line with short lengths of garden hose for quick filling. We don't use the original spouts because they are too slow. A drain in the floor carries drainage water out through a buried 4-inch plastic drain tile.

Past the sinks are two spinners—old washing machines—for drying greens, and behind the sinks is a 30-inch-by-6-foot stainless steel prep table

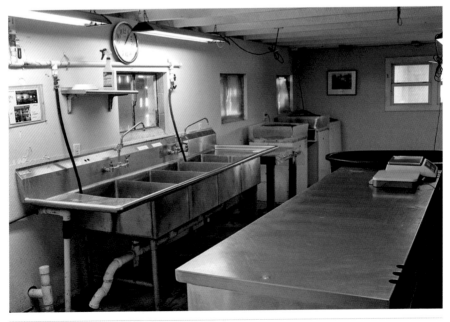

A processing room for greens washing should be easy to clean and well lit.

and an 18-inch foot-operated bag sealer. Open shelves behind the table hold packaging supplies and rags. We bought our sinks and the table from a bar going out of business and the bag sealer from the packaging company Uline. The washing machines, from a family member, were defective and we got them for free. We need only the spinning function.

Across the hall from the processing room is our magnetic whiteboard where we tally orders and use indicator magnets to show whether jobs are complete (for more on that system, see *The Lean Farm*). Under the board a table holds harvesting gear: nitrile-coated gloves, 24-by-½-inch produce twist ties for bunching, size 64 rubber bands for banding kale and head

Lean Rules for Keeping Work Areas Clean

1. *The short, high-frequency rule.* It is more efficient to clean in short spurts each day than in one big go at the end of a season. If you wait, you have suffered months of working in cluttered conditions, and stumbling around is a cost.

    We used to let the farm clutter up all spring, summer, and fall, throwing hoes into corners and letting dirt pile up on the floors. We thought (wrongly) that there were more important things to do than cleaning up. Now we practice short, high-frequency cleaning. We say: "Take it to zero." This means that after every job we aim to restore spaces to their original clean state—zero—as soon as practical. After bagging carrots, for instance, we put markers and labels away, sweep up debris, and give the floor a hosing. We are by no means perfect. We still lose knives and sometimes misplace tools. But this practice makes workdays more pleasant and frees up our winters for growing food and relaxing.

    To keep spaces at zero, hang a picture of perfectly clean spaces on the wall. With pictures, there is no need to explain to every worker what you mean by zero, and the whole crew stays motivated to match what they see.

2. *The one-speck-of-dust rule.* Our goal isn't actually to arrive at zero and stay there. It is to get cleaner every year. The *one-speck-of-dust rule* states that, at the end of the month or quarter or year, your farm should be just a tad cleaner than at the end of the previous time period. Trajectory is a powerful force. With this goal, you head in the direction of getting cleaner all the time.

    In the spring and fall we walk around and find some dirty window, barn corner, or floor to clean. In the spring of 2016, for instance, we washed the windows in the processing area, a job we had not done in several years. In the fall of 2016 we took everything out of the storage shed and swept out the corners. Routines like this keep our decluttering muscles strong.

lettuce, 4-by-4-inch sticker labels for wholesale bags, a roll of 2-inch painter's tape for labeling totes, and a cup of markers. That is all we need, and nothing else is allowed on the table.

2. SPRAY STATION

Your farm will also need an area about the size of a small garage for spraying crops such as carrots and radishes with water. An open-air lean-to suffices unless you grow early or late in the year, in which case you will want walls. Our spray station is a lean-to that we added to the barn, just outside the processing room. We installed two curtains, using leftover parts from a greenhouse project, which we roll up or down, depending on the weather.

In addition to hosing tables, the lean-to houses totes and crates for harvesting. When we built the lean-to we poured a cement slab that slopes to a drain basin in the corner. From there, water runs out through a 4-inch drain tile. We use two stainless shelves from a restaurant shelving unit as a spray surface. We used to support these shelves with crates functioning as table legs. Then an intern thought of a better idea: hang the tables with chains. Now we can quickly squeegee under them.

We hang both spray tables and hose in our spray station for quick clean up. The cement floor is easy to clean and slopes to a drain. Photo courtesy of David Johnson Photography.

We use salvaged electrical insulators to hang hoses in our processing area and in greenhouses.

Fourteen-gallon totes stored in a pyramid to dry. Notice that we keep them off the ground.

Another recent change was to hang the ¾-inch hose. This saves us from tripping, makes cleanup faster, and positions the hose ergonomically for washing. After consulting with a well service, we increased the diameter of the underground water line feeding the hose, from ¾ to 1¼ inches, to create high pressure. We use a brass shutoff valve as a spray nozzle. Since we spray totes and crates in the lean-to, we also store them there. The crates stack. We store the totes in a pyramid until dry, and then we collapse them.

3. COLD-STORAGE ROOMS

Many produce farms benefit from two cold-storage areas: one with temperatures around 55°F to 60°F, and one around 35°F to 40°F.

We call the 55–60°F room our tomato room. We use the room to box and store tomatoes and basil in the heat of summer, and to house storage vegetables in the fall. The room is 6 by 11 feet and is cooled with a standard window air conditioner. To set up a 35–40°F room, we bought a used 8-by-8-foot walk-in cooler box from a restaurant salvage business. We cut a hole in the back wall for an air conditioner that we regulate with a CoolBot external thermostat. The CoolBot-AC combination saves costs in two ways: the combination is much cheaper than a standard walk-in cooler condenser and evaporator, and the AC uses less energy. An alternative to a walk-in cooler box is to build your own super-insulated room (CoolBot offers plans on their website). We use the cooler to store all produce except for tomatoes and basil. In the winter we plug in an electric heater to keep the room just above freezing.

For a small farm, these two small air-conditioned spaces suffice. As your farm grows you can always add on more rooms, with more temperature options for specific crops.

4. STORAGE SHED

Another essential building is the storage shed, again about the size of a small garage, for housing equipment and supplies. You do not need heat or insulation in this building.

To create ours we walled off a portion of an old chicken house. In it we store four pieces of equipment: our tractor and walk-behind tractor (these do not need warmer temperatures in the winter), the Jang seeder, and the paper pot transplanter. Two shelves hold tightly rolled row covers. Other shelves hold tarps, irrigation supplies, and trellising supplies. In one corner we set up a small repair bench for sharpening tools and fixing handles. Because we farm with few tools, the small space more than suffices.

A low-cost alternative is to use a part of your greenhouse to store these items, although around March every year you might become jealous of the space. Think of what you could grow!

5. GAS HOUSE

I recommend that all farms set aside a small shed for storing gas, oil, and other flammable liquids. These items smell, pollute the ground, stain floors, and of course can burst into flames. They deserve a room of their own, set apart. We have two 55-gallon drums—one for diesel and one for gas—that we keep in a building the size of an outhouse, along with rags, engine oil, and other tractor and small gas engine fluids. The building sits 20 feet away from any other and can be locked for safety (important with children) and security.

6. RED-TAG ROOM

We sort tools every spring and fall, placing those that do not add value—that are not a part of our growing system—into a red-tag room until the next regional consignment auction, when we will get rid of those items. Using a red-tag room is part of the 5S system described in the *The Lean Farm*. A primary goal of 5S is to surround yourself strictly with the items you use, because extra tools, parts, materials, and supplies merely cause distraction. The red-tag room helps us focus on our work by eliminating clutter. We can still go back and pull out an item if we wrongly placed it there, and some-times we let one or two auction cycles pass before definitively getting rid of a tool. We think of the room as a vacuum cleaner, constantly sucking waste from the property. We use a walled-off corner of our barn, about 12 by 12 feet. We also set up a red-tag closet in our house. Just having such a space encourages us to fill it.

The Lean Method of Sorting

The organizing guru Mari Kondo says that when organizing a house, you should pick up each plate or book or shirt and ask if it "sparks joy." Don't overthink it. If there's no joy, get rid of the item. The effect is powerfully liberating. On our farm, we do a version of the same. Every six months we scour the farm looking for items that we did not use in the previous six months. We ask of every tool, "What task did you perform last season to add value to our farm?" If we struggle to find an answer, we remove the item to the red-tag room.

Lean practitioners use the word *seiri*, "sort," for this activity. Taiichi Ohno explains: "Sort means to throw out what you do not need, as when you do personnel adjustment. If you are holding on to your parts and stacking them up in your warehouse just because you worked hard to make them, this is not sorting."[1] When we started using lean, we sorted in one big go. We spent a few weeks filling trailers with old implements, scraps of fabric, rusty hoes—anything that did not fit our system—and sent them off to the auction. Our farm felt so light we almost danced a jig. We continue to practice *seiri* as part of our regular work.

7. IRRIGATION SYSTEM

A serious market garden requires a high-flow source of water. On rented property we used a portable gas pump to irrigate from a nearby canal. A 3-inch hose with a plastic head sat in the canal. Water passed through a sand filter system to keep small particulate matter out of the lines. It is important to place the pump close to the water source, rather than close to the garden, as pumps do a better job of pushing than pulling.

On our current farm we rely on a 4-inch well that supplies both our house and the farm. We use two pressure tanks—30-gallon and 60-gallon—for plenty of capacity. We run 1¼-inch water lines 4 feet underground to our hydrants. The frost-free hydrants feature 1½-inch-diameter pipes, wider than standard ¾- or 1-inch pipes in

A quick way to unroll drip tape is to pin the roll to the ground.

order to increase flow. The water is kept at a pressure of near 60 psi (pounds per square inch). Many online charts can help you choose well diameter, pump, and tank size, and water line diameter based on your needs, though I recommend consulting with a professional when designing your own system.

We use two types of irrigation: drip tape and overhead sprinklers. We use drip tape, with pre-punched emitter holes, on crops we grow under plastic or landscape fabric, including kale, tomatoes, peppers, zucchini, and cucumbers. The drip tape provides consistent moisture, uses less water, and keeps leaves dry, which is especially important for tomatoes. Our drip tape parts consist of a 2-inch header line, sufficient for long runs; shutoff valves; and 8 mm drip tape. We do not use caps on the end of the drip tape. Instead we tie them off by hand, using one or two simple knots with the tape itself. We use drip lines until they get plugged up with mineral deposits, usually after two seasons. To save lines for later use, we wrap them tightly and store them in tubs. We repair leaks with in-line couplers, but if there are more than two leaks we replace the line.

We use overhead watering as much as possible because it is lower cost and faster to set up than drip tape. We connect risers with ¾-inch hoses cut to length.

So that they move quickly, we use step spike bases from Orbitz on our irrigation risers.

We use overhead sprinklers, simple farm-built risers with Senninger Xcel-Wobbler heads, on crops grown in bare ground (not in plastic or landscape fabric) because they are faster to set up than drip tape. The heads will support interchangeable nozzles, depending on droplet size desired. We use size 10 turquoise, a medium size. In our tests, larger-diameter nozzles produced big droplets that displaced seed; droplets from smaller nozzles drifted away in the wind.

To make a riser, start with a step spike base for a garden hose and sprinkler, such as those made by Orbitz. Then use ¾-inch PVC pipe and fittings to assemble a riser. Ours are 4 feet tall, slightly higher than the tallest crops we plan to irrigate. Finally, screw on the Xcel-Wobbler heads. In winters, we remove the heads and store them in a bucket with a lid to keep spiders from building nests in the orifices. To connect hydrants and risers, we use ¾-inch commercial-duty hoses that lie on the inside edge of our plots. We connect the hoses to the hydrants using quick-connect fittings. An option for longer runs is to use poly tubing, which is stiffer to move about but cheaper.

The final piece of our irrigation system is timers. We use two types, both professional-duty, by Galcon. The first is a simple two-dial timer for general maintenance irrigating. One dial lets us choose duration—2 minutes to nine hours—and the other lets us choose how often—every three hours to every 14 days. We can change the settings in a matter of seconds. Because there is no LCD screen, the timer is durable. The second timer is a daily or weekly programming timer, which allows for more precise programming, though it is not as simple to use. We reserve it for germinating seeds when we direct-seed.

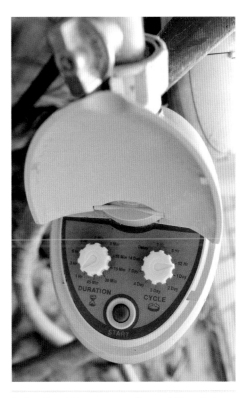

Irrigation timers, such as this Galcon two-dial timer, ensure that crops are watered on schedule.

8. VEHICLE ACCESS LANES

A final piece of infrastructure, often overlooked, is vehicle access lanes. You do not need polished concrete but you do need access to your growing plots and buildings. We built a lane running in front of our greenhouses and connected to the processing area. We can access these areas any time of year by vehicle, and trucks can bring heavy loads of manure or compost on-site without getting stuck. To build the lanes we started with a 6-inch primary base of #2 (3- to 4-inch-diameter) limestone, topped with a 2-inch layer of #72 (less than 1-inch-diameter) limestone.

Laying Out the Growing Beds

We grow all crops in beds, not in long rows. Beds are blocks of ground designated for crops where there is no walking. There are three reasons to use beds. The first is less motion waste because you walk a shorter distance to seed, plant, tend, and harvest crops. Image a line of 4,000 heads of lettuce, each spaced 1 foot apart. That is 4,000 feet—⅘ mile—to traverse from one end to the other. Four rows of head lettuce in a bed reduces walking by a factor of 4.

For easy access to our crops, 6-foot-wide mowed lanes surround our plots, each of which contains about eight beds. We grow in beds, not in long rows, to economize space. In the field, beds are 30 inches across the top and aisles are 18 inches wide.

The second reason is to make the most of our investment in our soil. As discussed, we have amended every square foot of land with costly compost. Planting in beds, rather than rows, means most of our land is used for food and very little for pathways.

The third reason for beds is to reduce compaction. When we limit walking to narrow pathways, the ground immediately around plants stays loose. That matters especially with young plants, which require loose soil at the root zone. Also, when you step around plants you always risk stepping on them or cracking stalks.

We have divided our farm into several plots, each roughly 30 feet wide by 65 feet long. Each plot holds around eight beds. We never need to cross more than four beds to reach a crop. Plots with a lot of beds in them add time to harvest because of the need to walk or drive around a larger area.

We chose a length of 65 feet for practical as well as psychological reasons. The practical reason: This was the distance available between the ends of our greenhouses and the pasture fences. The psychological reason: We never wanted to pick a row of beans longer than that. It was just too daunting. We would rather pick several rows 65 feet long rather than one row several hundred feet long. Now that we use the paper pot system for so many crops, we plan to adjust the bed lengths to match the length of our paper chains.

Our 6-foot-wide lanes are sufficient for carts and our Gator. They also provide runoff prevention, stopping topsoil from washing off the farm. At the ends of the beds we reserve 14 feet for comfortable tractor turnaround.

The beds themselves are 30 inches across the top, easy to straddle for harvest and to step across for quick movement across plots. The width also accommodates many tools designed for a 30-inch bed, including seeders and rakes. In reality, the width of beds is a personal matter. I have seen beds on successful farms that range from 24 to 60 inches wide.

Pathways between beds are 12 inches wide in the greenhouses and 18 inches in the field. Again, this is a personal matter, though there is no reason for pathways wider than you need in order to walk comfortably. We

How Close Is Too Close?

Based on our experience, these are minimum space requirements on vegetable farms with a small tractor:

- Driveway width: 8 feet
- Space between greenhouses: 12 feet
- Mowed lane between growing plots: 6 feet
- Clearance for turning a compact tractor between growing plot and buildings/fences: 14 feet

leave 12 inches around the perimeters of outdoor plots to allow for weekly shallow cultivating.

In our experience, the orientation of beds north, south, east, or west does not significantly affect plant growth, although on hilly ground it is best to run beds horizontally across slopes to help hinder runoff.

Putting It All Together:
The Lean Method of Laying Out a Farm

Laying out a lean vegetable farm starts with an analysis of motion. In lean factories managers trace the movements of their workers and design workspaces to minimize moves. They use a practice called *spaghetti diagramming*, in which one person is assigned to sit with notepad in hand and trace a line anytime a worker moves.

The result is a picture that often resembles a plate of spaghetti, and managers set about finding ways to *straighten*, *shorten*, and *eliminate* noodles. Noodles are straightened by minimizing awkward turns and twists. They are shortened by moving workspaces closer together. And they are eliminated when jobs are collapsed into one area, as when a plumber measures, cuts, solders, and pressure-tests a copper coil in one workstation instead of four. Farms can benefit from the practice. Consider the farm layout labeled "wrong."

I label the layout shown in the illustration "wrong" because, with functions spread out, the layout will involve a lot of wasted steps. You can easily measure this waste. I have put together a list of production steps to which I

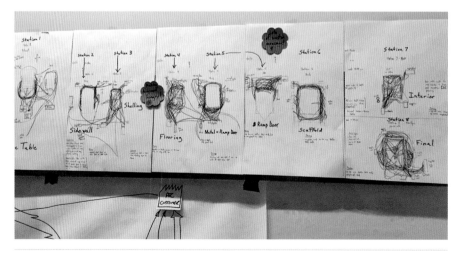

A spaghetti diagram of workstations at Aluminum Trailer Company, Nappanee, Indiana, used to design workspaces.

have added a number that represents a rough estimate of paces, or yards, a worker on a "wrong" farm might have to walk in order to grow and sell 100 heads of lettuce.

- Prepare soil (round trip from storage to field): 500
- Plant (same): 500
- Hoe (round trip from storage to field): 500
- Irrigate (distance from hydrant to field): 300
- Spray to ward off pests (same): 300
- Harvest (round trip from 40°F room to field): 700
- Wash (distance from 40°F room to processing room): 100
- Store in cooler (distance from processing room to 40°F room): 100
- Bunch or package (distance from 40°F room to processing room): 100

That totals *3,100* paces, which is a lot of moving simply because functions on the farm are far apart. Remember, you make money when you add value to your product, not when you walk around.

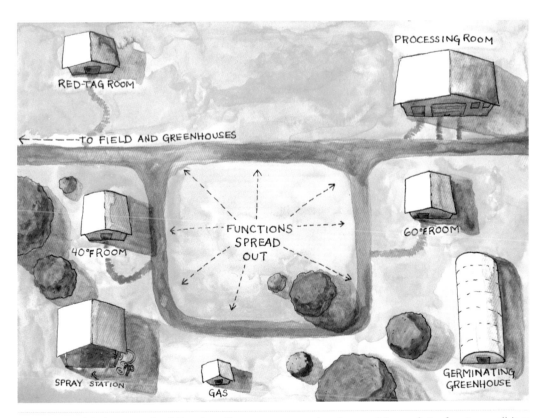

Wrong. In this layout all of the functions are spread out, requiring a lot of extra walking around, which results in wasted motion.

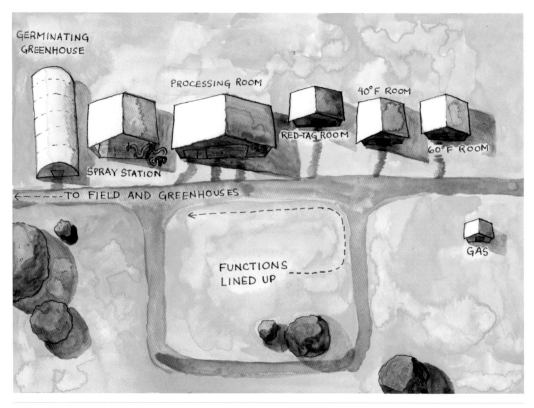

Better. In this layout, buildings are clustered and lined up. Lines of work are shorter and straighter. Buildings are placed in a logical sequence.

Now consider the farm design that I call "Better."

By clustering important functions, this layout *shortens* noodles, cutting the number of steps required for post-harvest processing. This layout also *straightens* noodles, reducing twists and turns needed to move from one station to the next. The building layout is sequential, following the order of work, which is easy to see: The germinating greenhouse lies on the road leading out to the fields; the spray station and processing room, followed by cold-storage rooms, sit waiting for harvested crops as they come in.

Let's go one step further. In the "best" layout, several of the functions—spray station, processing room, two cold-storage rooms, red-tag room, and germinating greenhouse—are compressed into one building, *eliminating* noodles altogether. That frees up space so that growing areas and hoophouses can be moved into the center of the farm. The number of steps required for 100 heads of lettuce with this arrangement is just 460, less than one-sixth of that required on the "wrong" farm.

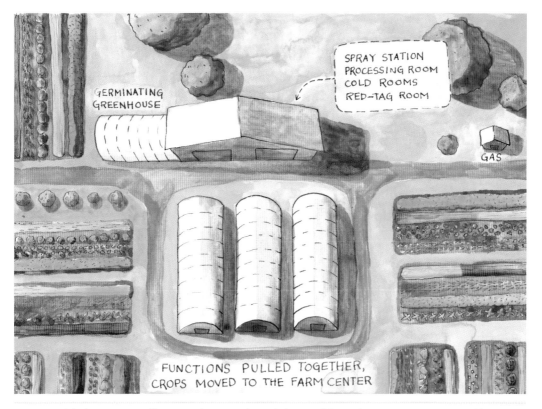

Best. In this layout noodles are short and straight, and functions are compressed into one building, eliminating noodles. Growing areas are at the center of the farm, not pushed to the edges.

- Prepare soil (round trip from storage to field): 100
- Plant (same): 100
- Hoe (round trip from storage to field): 100
- Irrigate (distance from hydrant to field): 30
- Spray to ward off pests (same): 30
- Harvest (round trip from 40°F room to field): 70
- Wash (distance from 40°F room to processing room): 10
- Store in cooler (distance from processing room to 40°F room): 10
- Bunch or package (distance from 40°F room to processing room): 10

When you lay out your own farm, stake the corners of buildings and growing plots and then walk between them. Pretend to carry an armful of transplants from germinating greenhouse to plots. Pretend to harvest salad greens, and then to hose, bag, cool, and send them out the door. Can you straighten, shorten, or eliminate noodles? Small farms have a key advantage over gigantic farms—the option of keeping workspaces close—so why not use it!

Some of the most exciting farm layouts I have seen bring buildings and functions together in extreme ways. One example is the German farm my grandmother grew up on, where animals lived in the basement of the house. Another model is the "connected farms" of nineteenth-century New England, where barns, sheds, and houses—the "big house," "little house," and "back house"—were connected, creating one enclosed farmyard.[2] This arrangement, considered progressive at the time, allowed for a sheltered south courtyard and enclosed access to all spaces in the winters. Also, of course, beautifully short lines of work.

CHAPTER 13

Leaning Up Greenhouses

Who loves a garden loves a greenhouse too.

—WILLIAM COWPER

I built my first greenhouse when I was a sophomore in high school, using scrap lumber lying around the family farm to erect a 10-by-12-foot A-frame. I covered it with a sheet of spare painter's plastic. I seeded spinach for our family to eat that fall and winter. I still remember standing in the humid, sunny greenhouse on a cold fall day, feeling like I had gotten away with a trick—as if I was in Florida while everyone around me was still stuck in Indiana.

Greenhouses are alluring structures and they deserve a place on the vegetable farm. Our farm would not be nearly as profitable without greenhouses. More than half of our sales are from greenhouses, and all of our high-dollar crops require them in order to serve our off-season markets. The primary advantage, from a lean perspective, is that greenhouses vastly reduce defects. Greenhouses are cozy places. Plants are protected from excess rain. The air is warmer. There is virtually no wind. With greenhouses, nearly every step we take counts because crops in these conditions are almost always successful.

A second advantage is that they extend the growing season, allowing a grower to deliver value to customers for a longer season of time. As a general rule, unheated greenhouses with one layer of plastic will move your farm one growing zone to the south (see appendix 5). Another layer of plastic and a bit of heat move you two or more zones south. Modern greenhouses allow for year-round production anywhere in the United States.

A third advantage is that greenhouses reduce cycle time, the time required to take crops from seed to harvest. Because the climate in a greenhouse is more controlled, crops grow faster—in some cases twice as quickly

Greenhouses are alluring structures. We look forward to spending sunny winter afternoons in ours.

as field-grown counterparts. For example, in our area, field-grown spinach seeded in October might be ready to harvest the next spring. In a greenhouse, we pick by Christmas. On small farms, where every square foot counts, that is a big deal.

We built one greenhouse per year on our farm until we had covered 9,000 square feet of ground with four structures, two of them heated and two unheated. The first permanent greenhouse we built, our propagation greenhouse, is the smallest, measuring 20 by 72 feet. Two layers of plastic cover the steel bows, with air blown between layers as insulation, a setup I would recommend with any heated greenhouse. The second, measuring 30 by 90 feet, also has two layers of plastic. Both are heated by Modine natural gas unit heaters, and are heated to 28°F in the fall and winter, and 60°F starting the first week of March, when we set out our first tomatoes. The remaining two are unheated, covered by a single layer of plastic. That is more than enough covered space to keep two workers busy all year.

From a lean point of view, the trick with greenhouses is to minimize the time spent building and managing them and maximize the time spent using them, even if this means more up-front costs. Value is built when you plant tomatoes and harvest peppers, not when you repair plastic or tear down a failed experiment. Here are my design principles for building a lean greenhouse, followed by three tips to trim *muda* from greenhouse management.

Greenhouse Terminology

I use *greenhouse* as a blanket term to refer to structures that protect plants and that are large enough to walk in. *Unheated greenhouse*, sometimes called a *hoophouse*, means the structure relies only on passive heating from the sun. *Heated greenhouse* means we add active heat with a heater, in our case natural-gas-powered unit heaters.

Design Principles for a Lean Greenhouse

An ideal greenhouse goes up quickly, uses as few parts as possible, is durable to withstand wind and snow, and requires minimal long-term maintenance. For beginners I recommend buying new, since you can consult with the manufacturer about your precise needs, and building will go much faster. As you gain experience, perhaps shop for used greenhouses or experiment with your own designs. Quality greenhouses are manufactured in every region of the United States. To reduce shipping costs, start your shopping close to home.

While having tried movable greenhouses, caterpillar tunnels, quick hoops, and other alternative forms of protected culture, we have come to prefer conventional, fixed-to-the-ground greenhouses because they have proved consistently reliable with a minimum amount of hassle, or what I call *muda* time (see "The Case for Permanent Greenhouses" sidebar on page 184).

Here are seven design rules to follow that are based on 10 years of experience with permanent greenhouses.

1. Bigger is better.
2. Sidewalls should be 5 feet tall.
3. Crossbraces should be 7½ feet tall.
4. Doors should be at least 8 feet wide.
5. A combination of manual and automated ventilation is best.
6. Build stiff endwalls.
7. Choose a site away from trees and close to home.

1. BIGGER IS BETTER

It is better to have fewer and bigger greenhouses than a lot of smaller greenhouses, for three reasons. First, like any other building, greenhouses require

The Case for Permanent Greenhouses

With permanent greenhouses, our focus is on growing food, not tinkering with buildings. Each of our greenhouses is cemented to the ground, with plastic securely attached by wiggle wire, a common fastening method. Because they are permanent, we do not spend time setting them up or tearing them down every year, and they require little maintenance. Permanent structures also allow us to install electrical systems required by our fans, louvers, thermostats, heating mats, germination chambers, and lights—all features that reduce *muda* and that are not always practical in temporary structures.

When we started farming, the first protected culture structures we used were low-cost hoops that we bent ourselves from 10-foot-long EMT (electrical metallic tubing), a type of steel electrical conduit, sometimes joining two pieces to create a structure we could walk under. These simple, temporary greenhouses allowed us to grow in protection with little investment, which was essential in the beginning when we were strapped for cash and growing on rented land. However, the hoops cost us dearly in time. Every season we devoted several days to building and deconstructing them. They also involved a lot of maintenance during the growing season to keep the plastic pinned down during winds and to ventilate when it was hot.

This is not to say you should never try an alternative greenhouse design. I am not opposed to creative solutions to the complex problem of protecting vegetables. Be realistic, however, about the amount of time such a structure will consume to manage compared with conventional greenhouses. Maintenance adds *muda*. Time spent seeding and harvesting adds value.

A downside of permanent greenhouses is that soil diseases and salt levels can build up. When greenhouses are moved or are taken down, rain and di-

maintenance. You need to replace plastic, fix curtains and doors, and mow around them. With 15 small structures, you will spend a lot of time running around them. Second, temperatures remain steadier in larger spaces because, with a larger air mass, space heats and cools more evenly. While we appreciate the smaller size of our heavily heated propagation greenhouse because we can save on heating costs, our larger greenhouses provide more stable long-season growing environments. Third is construction cost. Much of the cost of a greenhouse lies in the endwalls, with their fans, louvers, doors, polycarbonate sheets, and framing. The middle part of greenhouses is relatively cheap—just a sheet or two of plastic and bent pipe. While we built our four greenhouses one by one as our budget allowed, in retrospect one bigger greenhouse would have been easier to manage, at a lower total cost.

This permanent greenhouse features polycarbonate endwalls, an exhaust fan, automated louvers, 110-volt outlets, a hanging micro-irrigation system, and 5-foot-tall roll-up curtains—all features that shave off *muda* and that are not always practical in temporary structures.

rect sun help to mitigate these problems. While we have not yet needed to do so, our plan is to uncover each greenhouse for a few months every few years, and to periodically replace the top layer of soil, using our skid loader. That is a small price to pay considering the benefits of a fixed greenhouse.

The maximum size of a greenhouse depends both on snow loads, since wide structures are vulnerable to collapse, and on maximum temperatures in your region, since long structures develop hot spots in the middle. Growers in our area generally build as wide as 34 feet and as long as 148 feet, although there are exceptions. Your greenhouse manufacturer can give you recommendations for your own area.

2. SIDEWALLS SHOULD BE 5 FEET TALL

One of the design improvements we made with our latest greenhouse was to raise the sidewall from the standard 4 feet to 5 feet. At 5 feet we can easily work along the edge without awkward motion. Plants have more headroom

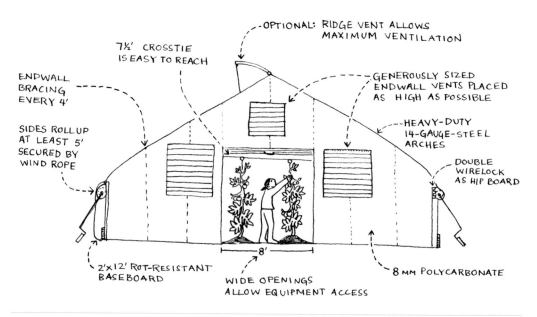

A greenhouse that serves the farm well uses minimal parts, is durable, and requires little maintenance. This illustration shows features of a well-built greenhouse.

too, since we can trellis crops on the outer rows instead of just in the middle. The 5-foot walls also allow for better ventilation via roll-up curtains.

We prefer straight rather than curved sidewalls. With curved sidewalls, if sidewall curtains are rolled up, rain from the roof runs off into the greenhouse. With straight sidewalls, rain only enters the greenhouse if the wind is blowing.

3. CROSSBRACES SHOULD BE 7½ FEET TALL

Crossbraces extend across a greenhouse to connect greenhouse arches together. Crossbraces are required if you plan to grow trellising crops such as indeterminate tomatoes or cucumbers. The crossbraces not only add strength but also provide handy framing for stringing wires and hanging twine. An ideal height for crossbraces is 7½ feet because that height allows them to be reached from tiptoe position (for most growers), or by standing on a bucket or small stool. In any given week we reach up to our crosses dozens of times to prune, add twine, or harvest.

4. DOORS SHOULD BE AT LEAST 8 FEET WIDE

We made the mistake in our first greenhouses of undersizing our doors. Wide doors allow easy tractor access. Also, on mild days we use doors to ventilate in lieu of rolling up sidewall curtains. On the east, west, and south sides, doors should be clear to allow sunlight to pass through. On the north side, you can use roll-up doors or other opaque options. All of our greenhouses have 8-foot-wide openings in the middle of each endwall plus one 32-inch-wide door for quick access.

5. A COMBINATION OF MANUAL AND AUTOMATED VENTILATION IS BEST

You almost cannot ventilate a greenhouse enough. I recommend a combination of manually operated sidewall curtains, also called roll-up curtains, and automated vents, typically in the form of aluminum louvers. Roll-up curtains allow a large amount of fresh air into the structure, rustling plants to stiffen them and dry them out. Automated louvers give you schedule flexibility: You can escape the farm without needing to worry about plants overheating. We control our louvers with thermostats in the middle of the greenhouse set at plant height. We set louvers to open at around 75°F. We chose the largest louvers possible with our endwall design. In our location, while louvers prevent killing freezing temperatures, they do not provide sufficient ventilation on their own during the hottest parts of the year.

A more advanced ventilation option is to install motorized endwall fans that pull fresh air through the greenhouse. With their high cost, we have never purchased new fans but rather search for used ones at auctions. By now we have installed fans on all of our greenhouses.

Another option is to install ventilation running along the peak of the greenhouse. Though costly, peak ventilation is the most effective way to prevent overheating. In combination with roll-up curtains, peak vents stimulate a chimney effect, causing fresh air to be pulled in from ground level, resulting in stable temperatures throughout the greenhouse. Peak ventilation is so effective that in some areas a peak vent can replace the need for all other forms of ventilation. We plan to put an automated peak ventilation system on any new greenhouse we build.

6. BUILD STIFF ENDWALLS

Unless your property is well protected from wind, I recommend that endwalls be made of a stiff material, not greenhouse plastic. On a north wall, sheet metal or wood paneling will do fine. On other sides I recommend clear polycarbonate panels. Stiff walls will encourage wind to splay around the greenhouse. Plastic covered endwalls catch wind like a sail, putting stress on the greenhouse. A flimsy endwall is often to blame when a greenhouse is blown away.

Selecting Crops for the Greenhouse

By now we have tried nearly every crop on our farm in the greenhouses. But not all make sense to grow there. Greenhouses represent high-cost space; they should be reserved for high-dollar-value-per-square foot crops or those that can be grown in a quick rotation.

Our basic schedule is to grow baby greens—mostly spinach—exclusively in the fall and early winter, plus one bed of kale. Local customers will buy whatever we can grow of these crops, and they offer the highest return per square foot because we can harvest them multiple times. In January, in addition to greens, we start to seed a wider range of fast-growing cool-season crops, like radishes, bok choy, and Hakurei turnips, in our 28°F greenhouses. In March, in the unheated greenhouses, we transplant more kale, and also grow the fast-growing cool-season crops. We reserve summertime for tomatoes, basil, peppers, and a row of ginger. The first tomatoes go in the first week of March. This calendar is simple to follow and it pushes out high-value crops all year from our greenhouses. Crops we usually keep out of the greenhouses include slow-growing carrots, full-size onions, and potatoes, and sprawling crops such as squash, melons, and corn.

Also, in the case of heated greenhouses, polycarbonate saves costs because it insulates better and lasts longer. If sheet plastic endwalls are your only option, then the sheets should be well braced with framing members behind them.

7. CHOOSE A SITE AWAY FROM TREES AND CLOSE TO HOME

Build greenhouses on your best land, as far from trees and other buildings as you can. Shade on a greenhouse slows plant growth, like putting the brakes on a car. Evergreen trees especially can spell dark days in winter greenhouses. Also, keep greenhouses close to other work areas. We are in and out of our greenhouses more than any other structure on our farm, besides our house. When choosing placement, count your steps and stay close to home.

What about orientation? In general, in northern climates greenhouses positioned east–west will allow in more light in the winter, since the lower sun crosses the sky near the horizon, shining into the long south side of the greenhouse. On the other hand, in east–west greenhouses tall summer plants shade each other more than in north–south greenhouses. Our experience is that orientation is a minor consideration compared with proximity to your house.

In the North, Perimeter Insulation Adds Efficiency

In our first few winters using greenhouses, we found that we could grow crops all winter in our tunnels—except along the outer edges, which froze. To reclaim that space we installed 1-inch-thick by 24-inch-wide rigid insulation all the way around the outside perimeter of each greenhouse, tucked just under the baseboards. The soil is slightly graded so the insulation slopes away from the greenhouse. We covered the insulation with landscaping fabric and wood chips. Now, except for during the most extreme cold, we can grow up to the edge. The insulation offers two side benefits: it sheds water 24 inches away from the greenhouse, keeping the edges drier, and it controls edge grasses. No more weeds creeping in from outside!

We installed 2-foot-wide rigid insulation, covered by landscape fabric, around each of our greenhouses to control weeds, shed water, and keep soils on the edge warmer. The fabric in the picture will be covered with wood chips.

Smarter Greenhouse Management

In order for greenhouses to pay, they must be well managed, not just well built. While greenhouse management is a topic that can fill an entire book, here are three specific practices we have adopted to lean our greenhouse growing.

1. Harden off plants in the fall.
2. Ventilate, ventilate, ventilate.
3. With supplemental heat, remember average temperature is what counts.

1. HARDEN OFF PLANTS IN THE FALL

When we started using greenhouses, we closed them up as much as we could in the fall, on the theory that we wanted to trap as much as heat as possible to speed growth. The results, however, were weak plants that were killed easily on the first truly cold night. They had been too coddled.

Now we leave our greenhouses completely open going into the fall and early winter, rolling up sidewall curtains and propping louvers, until cold air threatens to freeze crops. At that point, we close up. This exposes plants gradually to ever-harsher weather. The technique—a form of hardening off—ensures that plants are much more likely to survive when deep winter cold sets in.

Using this method, we now grow spinach all winter in unheated greenhouses, and can grow lettuces until temperatures reach well below freezing. Bok choy, kale, and other fall greens also survive longer when toughened up. In a sense, the greenhouses are aboveground root cellars preserving our crops. Rather than say we grow crops in winter, it is more accurate to say that we grow them in the late fall and preserve them in the ground.

During midwinter, we also install row covers as an additional blanket over crops, floating the covers directly on the crops, as with outdoor production.

2. VENTILATE, VENTILATE, VENTILATE

A challenge with greenhouse growing anytime of year is excess moisture trapped inside by the plastic. This leads to condensation on leaves and eventually to foliar disease. The shorter days and low light levels of winter exacerbate the problem.

How does condensation work? When I put the question to Vern Grubinger, vegetable and berry specialist at the University of Vermont, he told me, "Warm air holds more moisture than cold air." In winter greenhouses, "Condensation may occur in the evening as air cools and moisture drops out." Since winter nights are long, that condensate sits around for quite a while. Another answer I learned in middle school science class: warm, moist air condenses on cold surfaces.

Whatever the source of condensation in the greenhouse, the remedy is air exchange, pushing wetter air out and pulling drier air in. For this reason, we prefer that our winter unheated greenhouses stay "leaky." We don't pin down the sidewall curtains—a small amount of air can creep in under them—and we allow for small leaks around doors. To create leaks, one grower I have talked to leaves a top louver open all winter. On mild afternoons we open the endwalls of the greenhouses to let in fresh air. Anytime a greenhouse leaks you lose heat, of course, but it is far worse to lose an entire crop to mold for lack of fresh air. In our heated greenhouses, where we have sealed all leaks, we promote circulation by manually opening doors or by turning on endwall fans a few times each week when the weather is warmest.

On sunny afternoons, in both heated and unheated greenhouses, we also remove row covers. Besides better airflow, there are other benefits:

1. Plants receive the maximum amount of sunlight, encouraging darker leaves.
2. The soil soaks up UV rays, causing it to warm. When covers are replaced at night, they trap the heat in, close to plants.
3. There is less risk of freezing. Condensed moisture left on plants going into below-freezing nights can turn into ice, killing or damaging plants. Uncovering encourages plants to dry off before going into night. For the same reason, it is important to keep row covers dry by removing them completely from the winter greenhouse when irrigating.

At a minimum, covers should come off at least once per week, even if there is no sun.

3. WITH SUPPLEMENTAL HEAT, AVERAGE TEMPERATURE COUNTS

When we first applied minimal heat to our winter greenhouses, our practice was to set thermostats to 40°F day and night, thinking that would keep plants safely above freezing while promoting their growth.

Now we keep the nighttime thermostats set to 28°F. Why? Plant growth is relative to average temperatures in a 24-hour period, not to highs and lows. With that in mind, we now turn down the heat at night when heating is expensive, and frequently turn the heat up during the day, when heating is cheaper.

For example, on a sunny morning with a forecasted high of 50°F, I might turn the thermostat from 28°F to 70°F. It might cost a lot to heat from 0° to 40°F, but it costs relatively little to heat from 50° to 70°F, especially if the sun is helping. The result? Our plants bask in ideal photosynthesizing temperatures all day long, and grow. Come late afternoon, I turn the thermostat back to 28°F. Even if we do this only one or two days per week, plant growth

Average temperatures, not highs and lows, are what really count for plant growth. To save costs and boost growth in winter, we set thermostats low at night, when heating is expensive, and turn them up on sunny days, when heating is much cheaper.

is much faster. Following this method, we have cut our winter heating bill considerably, compared with when we heated to 40°F night and day.

With relatively affordable natural gas, supplemental heating pays off. With propane or other more expensive fuel sources, the payoff will be more limited. Also, bear in mind that heat alone does not make plants grow. They need light, too. So boosting heat too much when days are cloudy is a waste of energy and money.[1]

Why 28°F? In our experience this temperature setting keeps crops alive without the use of row covers if plants have been properly hardened off. On very cold nights, when frigid air presses at the sides of the greenhouse, we might cover more sensitive crops such as lettuce with a layer of row cover. Otherwise, 28°F has proved to be a safe low. Some growers might set thermostats even lower, for example down to the low 20s, and permanently cover crops with row covers. Others might turn heaters on just to warm the air for harvest, or to drive off cold air on the bitterest winter nights. When the forecast is for temperatures to dip to negative 20°F, a heater can be a handy tool in reserve, as long as the costs to heat build value. Once again, what works on one farm might not for another, since fuel costs vary widely.

CONCLUSION

The *Kaizen* Farm

We've got to get ourselves back to the garden.

—Crosby, Stills, Nash & Young

If you think you've arrived, you're ready to be shown the door.

—Steve Forbes

Between 1670 and 1810, farming animals—cows, horses, and oxen—all but vanished from the Japanese countryside. For example, in the Owari Domain, in the 1670s oxen and horses numbered 12,986—or about 50 animals per 1,000 people. By 1820, the number of work animals was 4,200—or about 13 animals per 1,000 people, a staggering decline of almost 70 percent (see table C.1).[1]

The animals disappeared because of Japan's growing population. As the island filled up, humans cleared away forests, where animals typically grazed, and the feed supply became too costly.

The loss of a draft animal is a big deal, especially with thousands of new mouths to feed. In wet paddy rice cultivation, the dominant type of farming in Japan at the time, animals shaped terraces, plowed, and hauled raw materials such as underbrush from the forest and human waste from villages. They transported crops to market. To cultivate rice without the help of draft animals, after growing accustomed to them, must have been unthinkably hard.

Table C.1. Change in the Population of Humans and Work Animals in the Owari Domain, 1670s to 1820[2]

Population	1670s	1820	Percent Change
Humans	265,522	331,678	+24.9
Work animals (oxen and horses)	12,986	4,200	−67.7

Note: In premodern Japan, work animals nearly vanished, while the number of mouths to feed increased by almost 25 percent.

Naturally, you would expect a decline in yields, lower living standards, and a slow descent into poverty. And yet—incredibly—the data tells us something else: productivity actually *increased*. Rather than starvation and poverty, the decline of animals precipitated "improvements in food, clothing and shelter." According to the Japanese historian Akira Hayami:

> *In the case in [sic] the Suwa region on [sic] Hinano Province, the life expectancy at the end of the seventeenth century was 25 years, but it increased to 35 years at the beginning of the nineteenth century. . . . This would not have occurred in conditions of poverty.[3]*

While 35 years is a short life span by today's standards, an increase of 10 years in life expectancy was a huge leap forward by historic standards, only made possible by better food access.

How did these farmers do it? First, they worked hard. They put in extra hours to make up for the loss of these animals. The writer Malcolm Gladwell notes, "Throughout history, . . . the people who grow rice have always worked harder than almost any other kind of farmer."[4] Take away animals, and the burden on humans undoubtedly increased even more.

But there is more to it. The farmers also used creativity to adapt. They formed co-op work groups, called *yui*, to overcome labor shortages during peak harvests. They reengineered and miniaturized tools to fit a human scale. For example, they replaced plows designed for oxen with ergonomic human-powered spades, which in some cases were actually able to plow deeper and better, increasing yields.[5] According to Gladwell, rice farmers throughout history "improved their yields by becoming smarter, by being better managers of their own time, and by making better choices."[6] The premodern Japanese farmers, in short, reinvented their farming.

In fact, the farmers ignited what Hayami calls a 400-year *industrious* revolution, where efficiency, hard work, and always searching for a better way became deeply ingrained in the Japanese psyche. This approach stood in stark contrast with Western agriculture at the time. In North America farmers were improving yields by

The Japanese Minō plow, designed to be pulled by a human, allowed for increased precision. Many other tools from Japan's premodern era exemplified simple but precise and efficient technology. They resulted in high-yield farming from small plots of land.

Better process is the heart of lean.

mechanical means, through an *industrial* revolution: machines replaced hand threshers, and tractors—not humans—replaced horses. Where Japanese farmers in the 1700s faced shrinking access to farmable landmass, US farmers were literally plowing westward, opening up thousands of acres of new ground. This is not to say that early farmers in North America weren't smart. But because of circumstance, farmers in Japan developed a special, and perhaps unmatched, emphasis on *efficiency*.

Toyota's first workers—the descendants of these rice farmers—brought the industrious way of thinking to the assembly line. In the early days at Toyota, Japan had been heavily bombed during World War II, and Toyota's production was well behind that of General Motors and other automakers. But these workers did not accept limitations. Within a few decades, Toyota became the world's leading automobile manufacturer, posting consistently high profits and remarkable employee retention rates—in a highly cyclical industry.[7] The fuel that powered the Toyota Production System was not technology—new hydraulic presses, computers, or robots. It was better process through continuous improvement—*kaizen*.

Our goal is to never stop improving.

This book started with the story of our farm, of how Rachel and I grew our business quickly but worked hard, nearly erasing our reserve of grit. Then I explained how we dedicated ourselves to the lean system, and showed in detail how we have rooted out waste from our production timeline. We now work nearly half as many hours without a cut in pay.

Our work is not finished, however. Our continuing job is to find ever-better ways to grow food. We use *kaizen*, the practice of continuous improvement, to organize our improvement efforts. The focus of *kaizen* is *process*, analyzing your actions and steering toward those that add value.

We can learn two lessons about *kaizen* from the story about Japanese farmers. First, *improvement requires constant change*. We tend to think that ancient farming was static, that people grew rice or beans or millet or wheat the same way for thousands and thousands of years. But that was not true. Growing systems were constantly changing as people moved, weather patterns changed, new technologies emerged, populations increased or declined, and so on. The most successful farmers throughout history adapted swiftly, not slowly, to change.

The second lesson is that *limits are a powerful force for accelerating change*. In the West we tend to see limits—in the amount of land we own, the level of our technology, the number of staff we can afford—as constraints holding us back. In fact, limits inspire new thinking, efficiency, and progress. We have scaled back our farm from more than 3 acres to less than 1, and every time we get smaller, we grow more profitable. The limit is not a constraint—it is a catalyst.

Here are three specific *kaizen* routines for yearly improvement that we use on our farm.

1. *We declutter our space more every year.* Even if we think we have gotten rid of every unused item, we scour our property again every year for tools and supplies that do not add value, and get rid of them. Even if we can eliminate just a few more items, we will increase our farm's flow, and enhance our ability to focus.

2. *We discover in more detail what our customers want.* Even if we think we have our customers pegged—sure that we know precisely what they want, when they want it, and how much—we go back to them again and ask the three questions: What do they want? When do they want it? How much do they want? Increased precision results in happier customers and higher profits. Every year we return to our chefs, we re-interview our CSA customers, and we collect more feedback from farmers' market customers. The information guides our farming.

3. *We root out more* muda *from our process.* Hanging a poster of the 10 wastes in the middle of our farm's processing area keeps us committed to eliminating more *muda* every year. To ingrain *kaizen*, we solicit improvement ideas weekly. We ask our workers who sell at our farmers' market booth to send a market report email that includes at least one new improvement idea. We discuss the email on Monday mornings. Not only do workers have permission to think up new ideas but doing so is part of their job description.

As I hope this book shows, with persistence we have chiseled many wastes from our farm. We have winnowed our stash of tools down to the essential. We have remodeled our buildings and laid out the property for smooth flow. We have removed motion waste from bed preparation and defect waste from seed starting. We transplant with just a few motions, and we sell with lean metrics to keep us on track. We are by no means perfect, but since employing lean, our timeline has shrunk while our profits continue to grow.

Lean, however, is much more than an exercise in cost cutting. It is about a cultural shift so that the focus of each day is on actions that build value. Our food should get better every year, not cheaper.

After 10 years of growing vegetables we realize that there is no plateau. There will never be a perfect garden. There will always be steps to eliminate and ways to more precisely give customers what they want.

Your own farm is no different. With a lean mind-set, you can stay small yet become highly profitable. You do not have to practice continuous expansion every year to grow your business. Practice continuous improvement instead: Identify value and cut out *muda* every time you work. Then enjoy a clutter-free, value-focused, and profitable farm.

Tools That Add Value

Below is a list of tools and supplies mentioned in this book, along with some suppliers. These tools are the survivors that made it through six years of *seiri* (sorting) sessions. The list of suppliers is not meant to be comprehensive.

1. Organizing and planning
2. Propagating and transplanting
3. Direct-seeders
4. Hand tools
5. Large equipment
6. Weed and pest management
7. Irrigation
8. Trellising
9. Harvesting and packaging
10. Farmers' market display
11. Clothes for market growing

Nolt's Produce Supplies is our go-to source for many standard produce-growing supplies. We like the selection, quality, and low prices that they offer. It is Amish-owned and not an online business, though a PDF version of their catalog is available online. Orders need to be placed by phone or mail. Although based in Pennsylvania, the business uses a network of distributors in Amish communities across the Midwest. Ask for the distributor nearest your location to pick up supplies in person.

1. ORGANIZING AND PLANNING

Magnetic Whiteboard, Grid, and Bicolor Magnets

The 5S Store
www.the5sstore.com

Garden Markers and Painted Garden Labels

Johnny's Selected Seeds
www.johnnyseeds.com

Farm with Just a Few Tools

For thousands of years humans have grown vegetables with the tiniest number of tools: a sturdy blade to turn new ground and to cultivate weeds, occasionally a sharp knife to harvest, and perhaps a basket to place crops in. In the nineteenth century, Henry Thoreau, the philosopher of simple living and a part-time market gardener, wrote that his tools at Walden Pond amounted to a hoe (for 54 cents) and a crow fence (for 2 cents).[1] Market gardening today requires more than that, but why get carried away?

A practice we follow in order to keep tools few is "Replace. Don't accumulate." When a new tool comes onto the property, the tool it replaces must quickly go. For example, after we purchased the BCS 710, we sold the old BCS immediately. When we upgraded to the Kubota tractor, we listed our old Fold 8N the same day. Practices like these keep our farming simple.

2. PROPAGATING AND TRANSPLANTING

Plastic Plug Flats

Nolt's Produce Supplies
www.noltsproducesupplies.net

10-by-20-Inch Leak-Proof Trays (for Bottom Watering)

Greenhouse Megastore
1020 Tray Shallow White
www.greenhousemegastore.com
We use the shallower 1.3-inch-deep trays rather than standard 2-inch-deep trays because 2-inch support trays cause sagging in the middle.

Germination Chamber Components

THERMOSTAT

Farmtek
DuroStat Watertight Thermostat with Remote Sensor
www.farmtek.com

500-WATT HEATING ELEMENT

Local hardware store

RUSTPROOF PAN FOR HEATING WATER

Local machine shop
Ask for help in fabrication.

Compost-Based Potting Mixes

Vermont Compost Company
Fort Light (for germinating crops in paper pot chains and smaller-celled flats)
Fort V (for everything else)
www.vermontcompost.com

Dairy Doo
Seed Starter 101 (for germinating crops in paper pot chains and smaller-celled flats)
www.dairydoo.com

Paper Pot Transplant Systems

Johnny's Selected Seeds
Paperpot Transplanter
www.johnnyseeds.com
This transplanter is slightly more adjustable than the transplanter from Small Farm Works, LLC, in order to work in a wider variety of soil textures.

Small Farm Works, LLC
Paper Pot Transplanter HP10
www.smallfarmworks.com

This transplanter is the original single-row transplanter. It is simple in design and lightweight. We have used this transplanter for several seasons.

Both suppliers offer the same components except for the transplanter itself. Other components we use:

> *Metal Rods (CP-1)*
> *Metal Chain Pot Frame (CP30K)*
> *Plastic Dibbling Board (DB-1)*
> *Two-Plate Gravity Seeder (SE-1)*
> *Paper Chain Pots (CP-303, LP303-10, LP303-15)*

Vacuum Seeder

Berry Seeder Company

Berry Precision Seeder
(good for small-scale growers)
www.berryseeder.com

Upon request, the Berry Seeder Company can manufacture seed plates designed for use with the paper pot system for those who prefer vacuum seeders over the two-plate gravity seeder.

Compost Covers

CV Compost
ComposText Covers
www.cvcompost.com

Compostwerks
TopText Composting Fleece
www.compostwerks.com

3. DIRECT-SEEDERS

1-Row Seeder, Jang Seeder
(Model JP-1)
Mechanical Transplanter Company
Jang Seeder (Model JP-1)
www.mechanicaltransplanter.com

EarthWay
1001-B Precision Garden Seeder
www.earthway.com

Johnny's Selected Seeds
Jang Seeder (Model JP-1)
EarthWay Vegetable Seeder
www.johnnyseeds.com

6-Row Seeder
Johnny's Selected Seeds
Jang Seeder, Model JP-6
www.johnnyseeds.com

4. HAND TOOLS

Planting Dibbler
Clarington Forge
Planting Dibber
www.claringtonforge.com
This tool is alternatively called dibber, dibbler, and dibble.

30-Inch Bed Preparation Rake
Earth Tools

SHW 54886 Aluminum Seedbed/
Hay Rake
www.earthtools.com

Row Markers for Bed
Prep Rake
Local hardware store
½-inch PEX plumbing cut to a length
of 6 inches

The Lean Method for Storing Tools

The lean method for storing tools is to keep them as close as possible to their points of use, at eye level, and, weather permitting, out in the open. Every tool should have a home. It should either be in its home or in your hands at all times.

When we started farming we stored all of our tools in a toolshed according to type—shovels in one corner, rakes on the wall, knives in a basket by the door. The system made us appear organized, but it wasn't lean. We constantly walked back and forth to the toolshed, which was at the edge of our production area, to get what we needed. As part of our leaning we spread tools out across the farm, placing them as close as possible to where we use them. Now when we need a tool, it is usually within a few steps, not across the farm.

On each of our four greenhouses we hang tools so they are within easy reach. In the winter we move the tools inside the greenhouses where we also have hooks for them.

Half-Moon Pull Hoe

Earth Tools

DeWit D10 Half Moon/Swan Neck Hoe
www.earthtools.com

This is our favorite all-around hoe. Ergonomically designed, it is strong enough to use as a grubbing hoe yet narrow enough for delicate jobs.

Long-Handled Wire Weeder

Johnny's Selected Seeds

www.johnnyseeds.com

Tine Weeding Rake

Johnny's Selected Seeds

21 Inches Wide
www.johnnyseeds.com

This is the rake we use after each harvest of baby greens to prepare crops for successive harvests.

Compost Fork

Earth Tools

Schwartzwaldschmiede, SHW 53862
Long Handled Compost Fork (54-inch)

Schwartzwaldschmiede, SHW 54532
Short Handled Compost Fork (44-inch)
www.earthtools.com

We use this sturdy, lightweight tool to clean up debris when rotating beds and to spread compost.

Wheel Hoe

Johnny's Selected Seeds

Glaser Wheel Hoe
www.johnnyseeds.com

Broadfork

Johnny's Selected Seeds

Johnny's 727 Broadfork (27-inch)
www.johnnyseeds.com

3-Tooth Cultivator

Johnny's Selected Seeds

www.johnnyseeds.com

Tomato Pollination Wand

Hydro-Gardens, Inc.

VegiBee Garden Pollinator
www.hydro-gardens.com

5. LARGE EQUIPMENT

Compact Tractor

Local dealer

Kubota Compact Tractor, model L3400, 11-inch ag tires, 4WD with "xTra Power"
This tractor is no longer made. Check with your local dealer to find comparable models.

Flail Mower

Purchased used locally

Lawn Genie Pickup Mower by Mathews and Co.

This flail mower features a catch basin so that cuttings can be easily transported around the farm if desired. We purchased it used, as the model is no longer made.

Rotavator

Local Woods dealer

Woods GTC60-2 (60-inch)

Root Digger

Local machine shop

We designed our root digger with a local machinist to fit our tractor. The blade is 38 inches wide between the two vertical rods so that it fully spans a 30-inch bed. The tip is tapered, helping the tool to dig in and go deep. Shortening the third arm on the tractor adjusts the angle. The digger can be removed and replaced by disks, which we use to hill potatoes. See photo on page 32.

Jiffy Hitch on our chisel plow.

Chisel Plow, 4-Shank
Purchased used locally

Compact Bed Shaper
Nolt's Produce Supplies
Model R8448, Serial 2009
www.noltsproducesupplies.net
This shaper is designed for use with compact tractors. With practice, beds can be spaced as close as 18 inches apart with this model.

Utility Vehicle
Purchased used online
John Deere 4x2 Gator
When shopping for a utility vehicle, make sure it can haul the weight you plan to carry. Golf carts are commonly used for carting around tools but most will not support a heavy load of compost.

Skid Loader
Purchased used locally
Gehl 4625 with 44 Diesel Kubota Engine
I recommend a skid loader with at least 40 hp and a minimum 5-foot bucket to move compost. Commonly available used.

Tiller
Local Honda dealer
Honda FG110 Tiller

Walk-Behind Tractor
Local BCS dealer
BCS 710

Hitch System
Jiffy Hitch System
www.jiffyhitchsystems.com
The Jiffy Hitch system includes a receiver that mounts to the tractor and triangular quick hitches that permanently mount on each implement.

6. WEED AND PEST MANAGEMENT

Mist Blower
Local SOLO dealer
SOLO 451 Motorized Backpack
Mist Blower

Landscape Fabric
Nolt's Produce Supplies
4-foot and 6-foot widths
www.noltsproducesupplies.net

Fabric/Ground Staples
Nolt's Produce Supplies
8-by-1-inch ground staples
www.noltsproducesupplies.net
I recommend these staples because they are longer than standard staples from hardware stores.

Sandbags
Uline
www.uline.com

Row Cover

Nolt's Produce Supplies

Agribon AG-30 Row Cover

www.noltsproducesupplies.net

We have also used Dupont 5131 (Typar T-518) row covers, also sold through Nolt's Produce Supplies. The Dupont covers last longer but are stiff to move about. Also, their fibers stick to plants as the covers deteriorate. This might be tolerable for watermelons, but not for greens.

7. IRRIGATION

Drip Tape Parts: 2-Inch Header, Shut-Off Valves, Drip Tape

Nolt's Produce Supplies

www.noltsproducesupplies.net

DripWorks, Inc.

www.dripworks.com

Overhead Irrigation Riser Parts

Nolt's Produce Supplies

Xcel-Wobbler (WOBX), lavender insert

www.noltsproducesupplies.net

PVC Schedule 40 (¾ Inch by 4 Feet) and Fittings

Local hardware store

½-Inch Step Spike Base

Plumbers Stock

Orbitz 58197N

www.plumbersstock.com

Timer

DripWorks, Inc.

Galcon Dial Timer

Galcon LCD Timer

www.dripworks.com

Overhead Irrigation in Greenhouses

Nolt's Produce Supplies

NaanDan Jain

"Greenhouse Hanging Assembly"

www.noltsgreenhousesupplies.com

We order 24-inch-length greenhouse hanging assemblies. We insert them every 3 feet into 1-inch black poly tubing that runs the length of our greenhouses. The black poly tubing rests on top of crossties. We fasten tubing to crossties with zip ties. We use two runs of tubing in each 30-foot-wide greenhouse.

8. TRELLISING

Tomato Stakes and Twine

Nolt's Produce Supplies

72-Inch White Oak Tomato Stakes

Winmore Tomato Twine

www.noltsproducesupplies.net

Netting

Johnny's Selected Seeds

Hortonova FG—79 inch × 250 foot

www.johnnyseeds.com

Trellissing Supplies

Hydro-Gardens, Inc.

Rollerhook

¾-Inch Black Plant Clips

www.hydro-gardens.com

9. HARVESTING AND PACKAGING

14-Gallon Storage Tote

Local hardware store

Roughneck storage tote by Rubbermaid

Curved Shears

Johnny's Selected Seeds

Curved Grape and Tomato Shears
www.johnnyseeds.com

We use these to harvest peppers, tomatoes, figs, and cucumbers. The 2-inch curve at the tip lets you harvest without twisting your arm and wrist into an awkward position.

Harvest Knives

Johnny's Selected Seeds

Stainless Steel Produce Knife, 6-Inch Blade
www.johnnyseeds.com

Our preferred knife for harvesting greens.

Paring Knife

Victorinox Swiss Army Inc.

Swiss Classic Tomato and Sausage Knife
www.swissarmy.com

Our serrated knife for general-purpose harvesting.

Greens Harvester

Farmer's Friend

15-Inch Quick Cut Greens Harvester
farmersfriendllc.com

Digging Fork with D Handle

Clarington Forge

Garden Fork with "D" Handle (40-inch)
www.claringtonforge.com

Walk-In Cooler Controller

CoolBot

www.storeitcold.com

Allows a window air conditioner to function as a walk-in cooler compressor.

Nets for Processing Salad Greens

FNC Nets and More

N9-1045, #9 nylon netting (multifilament)
1-inch square, 2-inch strand, 45 mesh
www.netsandmore.com

We use these nets to move greens from one wash tank to another. First we make frames out of ¾-inch PVC pipe. The frames are slightly smaller than the tanks. As mentioned, we drill holes in the frames so that they sink. Next, we fasten 1-inch-square nylon fish netting to the pipes using zip ties, creating a sort of large pool skimmer without a handle. See photo on page 139.

Packing Boxes

Monte Package Company

www.montepkg.com

The most common packaging supplies we order are ½- and ¾-pound produce boxes (for CSA shares), and 10-pound tomato flat-pack cartons and 25-pound tomato cartons (kraft body).

Produce Twist Ties

Hubert Company

Package Containers White Paper
Produce Tie with Green Imprint
"Locally Grown"—18 inches L × ¾ inch W
www.hubert.com

We cut these in half when they arrive, and then band them in groups of 10 to make counting easier at harvesttime. We use them to bunch turnips, beets, carrots, and other roots that we sell with their tops on.

Rubber Bands

Amazon

Alliance Sterling, Size 64
www.amazon.com

We use these to band romaine, head lettuce, and kale.

Plastic Bags and Labels

Uline

S-3202: 9×12 inch, 1 Mil Poly Bag
(for greens sold in ¼-pound units)
S-3203: 10×14 inch, 1 Mil Poly Bag
(for greens sold in ½-pound units)
S-3204: 12×18 inch, 1 Mil Poly Bag
(for greens sold in 3-inch units)
S-10912: 26×36 inch, 1 Mil Poly Bag
(for packaging bulk-order crops for
wholesale buyers)

S-17069: 4×4 inch,
White Industrial Weatherproof
Thermal Transfer Labels—
Polypropelene (for labeling bags sold to
wholesale buyers)
www.uline.com

Sealer

Uline

H-89 Foot-Operated Impulse (18-inch)
www.uline.com

10. FARMERS' MARKET DISPLAY

Storage Baskets

Hubert Company

Black Wicker Storage Baskets
www.hubert.com

Clear Sign Clip

Hubert Company

44498: Clear-Plastic, Extended-Height,
Clip-On Signholder—6 inches High
www.hubert.com

These versatile clips allow us to quickly post our prices up in the air where customers can see them.

11. CLOTHES FOR MARKET GROWING

Footwear

Crocs Retail, LLC

Crocs Bistro
www.crocs.com

These are my preferred clogs for summer. Designed for chefs, the shoes feature a slip-resistant sole, adding safety in the processing room.

Birkenstock USA LP

Birkenstock Profi
www.birkenstockusa.com

I prefer this sturdier clog in colder weather and when the ground is wet.

Gloves

Amazon

Atlas Nitrile Tough Work Glove with
Nylon Liner
www.amazon.com

We use these to warm our hands for cold-weather harvests. We purchase a cut-resistant version that workers sometimes use when cutting baby greens.

Pants

Kühl

Kühl Revolvr
www.kuhl.com

These lightweight pants dry quickly and feature articulated knees designed for kneeling.

Clay Bottom Farm
Seed Varieties

Seed Varieties

Crop	Seed Varieties
Salad Greens	
Arugula, bagged	Arugula (regular)
Bok choy	Joi Choi
Microgreens	Mizuna, Red Rambo Radish, Hong Vit Radish, Red Russian Kale
Mizuna	Mizuna, Miz America, Red Kingdom
Pea shoots	Austrian Winter
Salad mix, bagged	Red Salad Bowl, Defender, Dane, tatsoi, Carlton, mizuna, Garrison
Spinach, bagged	Gazelle
Bunched/Bagged Roots	
Carrots, bunched or bagged	Yaya, Nelson, Bolero
Golden beets, bunched or bagged	Touchstone Gold
Green onions	Guardsman
Radishes, bunched	D'Avignon, Pink Beauty, Patricia, Rover
Red beets, bunched or bagged	Red Ace, Moneta
Turnips, bunched	Hakurei
Leafy Greens	
Curly kale	Winterbor
Lacinato kale	Black Magic
Multileaf head lettuce	Salanova® Red Sweet Crisp, Multigreen MT 3, Multired 4, Multired 54
Romaine	Green Forest, Coastal Star
Swiss chard	Bright Lights

Crop	Seed Varieties
Herbs	
Basil	Eleonora, Amethyst Improved
Cilantro	Santo
Parsley	Moss Curled II
Tomatoes	
Artisan tomatoes	Sunrise, Pink, Purple, Bumble Bee
Heritage tomatoes	Marnero, Margold, Pink Berkeley Tie Dye, Green Zebra
Red or yellow tomatoes	Carolina Gold, BNH 589, BHN 871, Glacier
Peppers	
Colored pepper mix	X3R® Red Knight, Cheyenne, Moonset, Sprinter, Islander, Purple Beauty, Early Sunsation, Milena
Other	
Asparagus	Jersey Supreme
Beans	Provider, Royal Burgundy, Rocdor
Bulb fennel	Orazio
Cucumbers	Corinto
Edamame	Butterbean
Eggplant	Orient Express
Garlic	Local variety and Music
Kohlrabi	Winner
Leeks	Megaton
Pink ginger	Big Kahuna
Rhubarb	Canadian Red
Shallots	Conservor
Sugar snap peas	Super Sugar Snap
Zucchini	Dunja

Dollar-Value-per-Bed Chart

T   *he Lean Farm* includes a chart that provides the dollar value per square foot of selected crops that we grow. Here I calculate those values per 30-inch-by-50-foot bed (125 square feet), a standard bed size on many farms. Many conditions can lower yield, and of course prices vary regionally, so do not expect replicate results. Figures for tomatoes, greens, and ginger are based on yields from crops grown in greenhouses. All figures are based on the average yield of at least two harvests. While we focus on crops higher on the list, we still sometimes grow small amounts of those that score lower to add diversity to CSA boxes, or we purchase those items from other growers. We are transparent with our customers about such arrangements.

Table 17.1. Approximate Dollar Value per 30-Inch-by-50-Foot Bed: Selected Crops at Clay Bottom Farm

Heirloom tomato	$2,406.00	Rhubarb	468.75
Hybrid tomato	1,500.00	Turnips, bunched	412.50
Ginger	1,500.00	Garlic	375.00
Pea shoots	1,250.00	Red beets, bunched	350.00
Salad mix	1,250.00	Fennel, bulb	350.00
Spinach	1,250.00	Kohlrabi	350.00
Spring mint tips	937.50	Head lettuce	312.50
Romaine	625.00	Green onion	312.50
Carrots, bunched	562.50	Bok choy	312.50
Carrots, bagged	562.50	Potatoes, new	162.50
Shallots	562.50	Broccoli	156.25
Microgreens	468.75	Onions	125.00

Photo Gallery of Tips for Seven More Crops

This photo gallery presents lean thinking applied to seven more crops that we grow.

RHUBARB

Rhubarb is a high-profit-margin crop because it grows abundantly with little care. The key to large rhubarb stalks is to fertilize in the fall with 1 inch of high-N compost. We space plants 3 feet apart in rows 4 feet apart with landscape fabric between rows.

Our preferred variety is Canadian Red. It is slow to bolt and boasts a bright red stalk.

We cut rhubarb in the field, leaving a few inches of greens for looks.

GARLIC

We grow garlic in raised beds, using cloves stored from the previous year. We shallowly till in a 1-inch layer of high-N compost before planting, which occurs between September 15 and October 15. We space seeds four rows per 30-inch bed, 12 inches apart in the row.

Contrary to common practice, which is to cure garlic by hanging the entire plant, we cut garlic heads in the field, leaving a few inches of stem as a handle. This leaves the mess in the field and allows for curing in a small space. Garlic is ready to harvest when all but the bottom three leaves have turned brown.

We cover garlic with a thin layer of straw or hay (mostly grass) mulch. This step, which is optional, helps hold in moisture. The key to large heads, in addition to the high-N compost, is to provide consistent water, especially in the month before harvest.

CUCUMBERS

We cure garlic, along with other alliums, in an insulated room. A dehumidifier sucks moisture from the heads and a high-velocity fan (just outside the picture) circulates air. The process takes two to three weeks. When stems feel dry we move them to a cool, dry room and sell throughout the fall and early winter. We sell hardneck varieties first and save softneck varieties, which store longest, for last.

We hose garlic clean before curing. Because we use a controlled curing method (see above), the water does not damage stored bulbs.

Cucumbers are an easy-to-grow crop, provided we fertilize them well and protect them from cucumber beetles. Our favorites are thinner-skinned varieties. We amend their soil with a 1:1 ratio of high-N and low-N compost and grow them under row cover to exclude pests until they start to flower. Like ginger, they require consistent soil moisture, which we provide through drip tape.

Cucumbers can sprawl on the ground, but for straight fruits, which sell best in our markets, we trellis them. At first we grew them up strings in the greenhouse, using an umbrella trellising method commonly used by greenhouse cucumber growers. However, because of their relatively low price point in our markets, we now grow them in the field on Hortonova FG netting, a trellis designed for climbing vegetables. This encourages straight fruits and saves space (see photo on page 114). We space plants 12 inches apart, one row per 30-inch bed, and train them up the netting by hand, with the occasional help of tomato vine clips, with no pruning. This method provides abundant harvests for our markets with much less work.

SHALLOTS

Shallots make an ideal paper pot crop because they are time consuming to transplant by hand, yet sell at a high price per pound. We start them in mid-February, in germination chambers set to 78°F. As they grow in the paper pots we cut their tips to 3 inches every few weeks to promote stocky transplants.

To prepare beds for shallots we mix in a thin layer of high-N compost prior to transplanting. We grow them three rows per 30-inch bed, which leaves room to side-dress them with high-N compost in midseason if needed.

We harvest, bunch, and clean fresh shallots in the field, using a single piece flow approach, and sell them three to four bulbs per bunch. If needed, we hose them in the processing area. Later in the season we sell them cured and in bulk, using the same curing method as for garlic.

LEEKS

Because leeks require a long growing season, we often transplant them into plastic, three rows per bed, 6 inches apart in the row. As with garlic and shallots, we blend in high-N compost prior to transplanting. We use drip tape so they receive consistent water. Overhead irrigation encourages rot late in the year.

We start leeks in February in 2-inch-deep open flats, five rows per flat with about ⅛ inch between seeds in the rows. We sprinkle the seeds in by hand and lightly cover them. We do not trim leeks, because we want tall seedlings. We add high-N compost to the seedlings if they look pale green.

We achieve white shanks, preferred by our customers, by transplanting leeks 9 inches deep, using the dibbler. We poke a hole, drop in a baby plant, and let dirt fill the hole through natural settling.

MICROGREENS

In *The Lean Farm* I describe how we shaved moves from our microgreens process. In short, our process involved spreading seeds by hand across shallow open flats. We now use paper pot tools. We use the dibbler and two-plate gravity seeder to seed paper pot support trays, without the paper chains. We aim for six to eight seeds per hole in the gravity seeder. This process spaces seeds more accurately than hand seeding. Consequently, flats grow more evenly, the greens receive better airflow, and we avoid clusters of seeds spaced too closely together, where mold often grows. This photo shows purple mizuna starting to sprout.

We grow pea shoots in the same manner, except we still spread them by hand, about 2 cups per flat. All microgreens flats start out in the germination chamber. We harvest the greens with 6-inch produce knives, the same knives we use to harvest baby greens.

Purple mizuna nearly ready for harvest. Our favorite crops for microgreens are radishes and mizuna because of their fast growth and low seed cost.

We wash microgreens and pea shoots in the same manner as greens (see chapter 10), except we spin them in mesh laundry bags.

GINGER

Big Kahuna, shown here, is our preferred ginger variety. From seed to harvest ginger prefers temperatures to be around 72°F. If you are comfortable, your ginger is comfortable. We start sprouting it in February or March in our mini-greenhouse (the greenhouse-within-the-greenhouse), burying pieces about an inch apart in 2-inch-deep 10-by-20-inch flats filled with peat moss. We keep the flats consistently moist. We avoid compost in the growing medium because it encourages rot. It is important to thoroughly pre-moisten the peat moss.

After sprouts reach about 6 inches tall, we grow the ginger just like potatoes, hilling them once per month to cover exposed tubers. Here they grow in a tomato greenhouse. Ginger likes filtered sunlight, which the tomatoes provide.

We amend their soil with 1 inch of high-N compost before placing them in the ground, and we side-dress plants midseason with high-N compost as well. We grow one row per 30-inch bed with seed pieces 12 inches apart in the row. It is important to keep the ground consistently moist.

USDA Zone Map

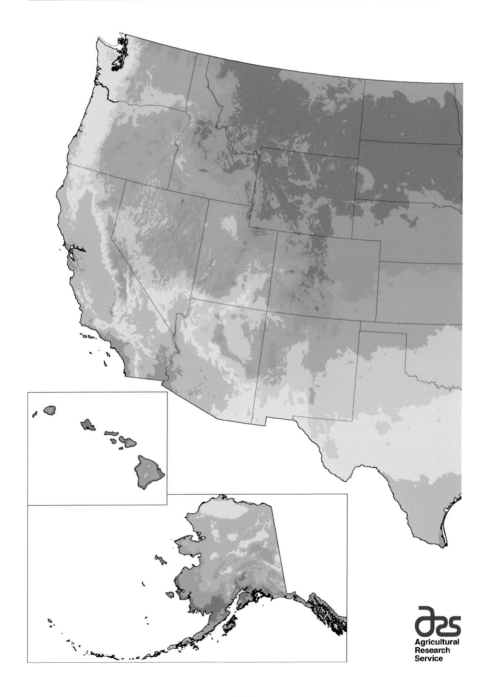

Agricultural
Research
Service

The seeding dates specified in this book correspond to to our zone, 5B. If you are in a different zone, adjust your seeding dates accordingly. The seed starting date calculator found at www.johnnyseeds.com/growers-library /seed-planting-schedule-calculator.html will aid in this task. For a more detailed Plant Hardiness Zone Map, see http://planthardiness.ars.usda.gov /PHZMWeb/Maps.aspx.

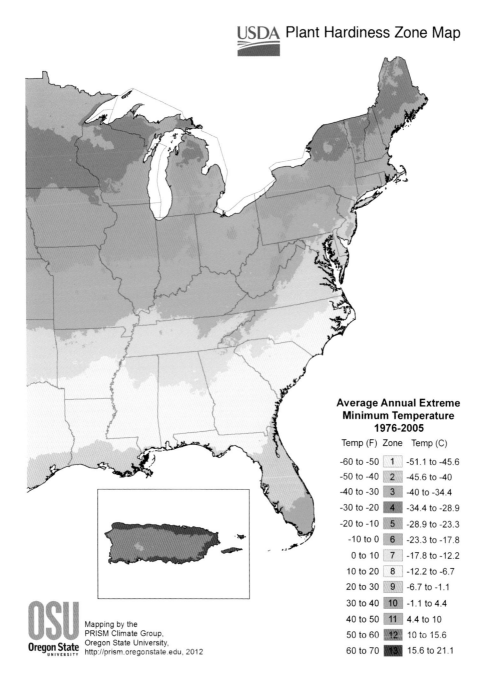

USDA Plant Hardiness Zone Map

Average Annual Extreme Minimum Temperature 1976-2005

Temp (F)	Zone	Temp (C)
-60 to -50	1	-51.1 to -45.6
-50 to -40	2	-45.6 to -40
-40 to -30	3	-40 to -34.4
-30 to -20	4	-34.4 to -28.9
-20 to -10	5	-28.9 to -23.3
-10 to 0	6	-23.3 to -17.8
0 to 10	7	-17.8 to -12.2
10 to 20	8	-12.2 to -6.7
20 to 30	9	-6.7 to -1.1
30 to 40	10	-1.1 to 4.4
40 to 50	11	4.4 to 10
50 to 60	12	10 to 15.6
60 to 70	13	15.6 to 21.1

OSU Oregon State UNIVERSITY

Mapping by the PRISM Climate Group, Oregon State University, http://prism.oregonstate.edu, 2012

Glossary of
Japanese Terms

Heijunka. Production leveling or "leveling the load" in order to avoid rushes, which overburden people and equipment, and down time, which underutilizes fixed costs like buildings and equipment. Lean managers practice *heijunka* in order to create smoother and more predictable work environments, on the theory that fluctuations in production always increase waste.

Kaizen. Continuous improvement. The goal of *kaizen* activities is to discover improvements and banish waste until a firm achieves zero-waste production. That goal might never be attained but it still provides inspiration to improve.

Kamishibai. A form of storytelling using scrolls and illustrations performed by twelfth-century Buddhist monks in Japan. Managers at Toyota use a version of the practice to instruct workers and to perform audits through visual systems. *Kamishibai* boards involve color-coded cards placed on a board.

Kanban. A replacement signal. For example, when milk was delivered in bottles, empty bottles set outside the door functioned as a *kanban*, a cue to replace. In lean operations, *kanbans* assist with just-in-time delivery of supplies and parts.

Muda. Waste in the form of steps that do not actually create value for customers.

Mura. Waste in the form of unevenness in production and sales, which creates erratic, unpredictable work environments.

Muri. The waste of overburdening people or equipment. *Muri* leads to equipment failure, poor work, injury, and burnout.

Seiri. To sort. In lean, *seiri* means getting rid of everything not absolutely necessary for current production. The first step of the 5S organization system.

Seiketsu. To standardize and make part of everyday work the practices of 5S. The fourth step of the 5S organization system.

Resources
for Further Study

I recommend the resources below, alphabetized by title, for those who want to delve deeper into concepts and practices in this book.

RECOMMENDED READING ON ORGANIZING AND SIMPLIFYING

The Life-Changing Magic of Tidying Up: The Japanese Art of Decluttering and Organizing, by Marie Kondō (Ten Speed Press, 2014).

The Organized Mind: Thinking Straight in the Age of Information Overload, by Daniel J. Levitin (Dutton, 2015).

Spark Joy: An Illustrated Master Class on the Art of Organizing and Tidying Up, by Marie Kondō (Ten Speed Press, 2016).

RECOMMENDED READING ON SMALL-FARMING PRACTICES

Compact Farms: 15 Proven Plans for Market Farms on 5 Acres or Less, by Josh Volk (Storey Publishing, 2017).

The Greenhouse and Hoophouse Grower's Handbook: Organic Vegetable Production Using Protected Culture, by Andrew Mefferd (Chelsea Green Publishing, 2017).

Growing for Market. A magazine for market growers.

"Insecticides Comparison Chart," by Johnny's Selected Seeds (2016), www.johnnyseeds.com/growers-library/tools-supplies/insecticides -comparison-chart.html. We refer to this chart when we confront a pest problem we can't control through exclusion.

The Market Gardener: A Successful Grower's Handbook for Small-Scale Organic Farming, by Jean-Martin Fortier (New Society Publishers, 2014).

"Soil Temperature Conditions for Vegetable Seed Germination," by Oregon State University Extension, http://extension.oregonstate.edu/deschutes /sites/default/files/Horticulture/documents/soiltemps.pdf. We use this guide to help determine temperature settings for seed germinating.

Sustainable Vegetable Production from Start-Up to Market, by Vernon P. Grubinger (Natural Resource, Agriculture, and Engineering Service, 1999).

The Urban Farmer: Growing Food for Profit on Leased and Borrowed Land, by Curtis Stone (New Society Publishers, 2015).

"Web Soil Survey," by United States Department of Agriculture, websoilsurvey.sc.egov.usda.gov/App/HomePage.htm.

The Winter Harvest Handbook: Year-Round Vegetable Production Using Deep-Organic Techniques and Unheated Greenhouses, by Eliot Coleman (Chelsea Green Publishing, 2009).

RECOMMENDED READING: FARMING CLASSICS

Farmers of Forty Centuries: Organic Farming in China, Korea, and Japan, by F. H. King (Dover Publications, 2004).

Five Acres and Independence: A Handbook for Small Farm Management, by M. G. Kains (Dover Publications, 1973).

Knott's Handbook for Vegetable Growers, 3rd ed., by Oscar A. Lorenz and Donald N. Maynard (John Wiley, 1988).

The Organic Front, by J. I. Rodale (Rodale Press, 1949).

COMPOSTING RESOURCES

"Compost Calculations," by Katie Campbell-Nelson (University of Massachusetts Extension, September 2015): https://ag.umass.edu/sites/ag .umass.edu/files/projects/related-files/compost_calculations.pdf. Use this guide to calculate compost nutrients either by volume or by weight.

"Composting on Organic Farms," by Keith R. Baldwin and Jackie T. Greenfield (North Carolina Cooperative Extension Service, Center for Environmental Farming Systems, 2006): content.ces.ncsu.edu /composting-on-organic-farms. Provides a detailed C:N chart of various compost ingredients.

"Nutrient Management for Fruit and Vegetable Crop Production: Using Manure and Compost as Nutrient Sources for Vegetable Crops," by Carl J. Rosen and Peter M. Bierman (University of Minnesota Extension Service, 2005): http://www.extension.umn.edu/garden/fruit -vegetable/using-manure-and-compost/docs/manure-and-compost .pdf. This is a seven-step guide to develop a finely tuned composting plan for initial soil buildup.

"Soil Fertility Management for Organic Crops," by Mark Gaskell, Richard Smith, Jeff Mitchell, et al. (University of California–Davis, Vegetable Research and Information Center, 2007): http://anrcatalog.ucanr.edu

/pdf/7249.pdf. Provides an excellent overview of organic soil fertility, including through the use of compost.

"Soil Organic Amendments: How Much Is Enough?," by Thomas F. Morris, Jianli Ping, and Robert Durgy (University of Connecticut): http://www.newenglandvfc.org/pdf_proceedings/SoilOrganicAmend.pdf. Provides an in-depth discussion of excess P and N in soil.

HISTORY OF FARMING IN JAPAN

Can Japanese Agriculture Survive? A Historical and Comparative Approach, by Takekazu Ogura (Agricultural Policy Research Center, 1979).

The I Ching in Tokugawa Thought and Culture: Asian Interactions and Comparisons, by Wai-Ming Ng (University of Hawaii Press, 2000).

Japan's Industrious Revolution: Economic and Social Transformations in the Early Modern Period, by Akira Hayami (Springer Japan, 2015).

The Making of Modern Japan, by Marius B. Jansen (Harvard University Press, 2000).

Notes

PREFACE

1. F. H. King, *Farmers of Forty Centuries: Organic Farmers in China, Korea, and Japan* (Mineola, NY: Dover Publications, 2004), 9.
2. King, *Farmers of Forty Centuries*, 19.

INTRODUCTION: FIVE STEPS TO A LEAN VEGETABLE FARM

1. Eliot Coleman, *Four Season Harvest: Organic Vegetables from Your Home Garden All Year Long* (White River Junction, VT: Chelsea Green Publishing, 1999), 14–15.
2. James P. Womack and Daniel T. Jones, *Lean Thinking: Banish Waste and Create Wealth in Your Corporation* (New York: Free Press, 2003), 323.
3. Taiichi Ohno, *Toyota Production System: Beyond Large-Scale Production* (Portland, OR: Productivity Press, 1988), ix.

CHAPTER 1: PLANNING THE YEAR WITH *HEIJUNKA* AND *KANBAN*

1. Natasha Bowens, "CSA Is Rooted in Black History," *Mother Earth News*, February 13, 2015, http://www.motherearthnews.com/organic-gardening/csas-rooted-in-black-history-zbcz1502.
2. Simon Huntley, "CSA: We Have a Problem," Small Farm Central, August 17, 2016, http://www.smallfarmcentral.com/Blog-Item-CSA-We-have-a-Problem.
3. Jim Womack, coauthor of *Lean Thinking*, introduced the concept of emotional *heijunka* to me in an email, July 10, 2016.

CHAPTER 3: COMPOST MAKING, SMALL FARM–STYLE

1. Charlotte von Verschuer, *Rice, Agriculture, and the Food Supply in Premodern Japan* (London: Routledge, 2016), 17–19.
2. F. H. King, *Farmers of Forty Centuries: Organic Farmers in China, Korea, and Japan* (Mineola, NY: Dover Publications, 1911), 47.

3. King, *Farmers of Forty Centuries*, 73.

4. King, *Farmers of Forty Centuries*, 208–209.

5. King, *Farmers of Forty Centuries*, 11.

6. US Composting Council, "Persistent Herbicide FAQ," accessed March 22, 2017, http://compostingcouncil.org/persistent-herbicide-faq.

7. Steven Wisbaum, "In Defense of the Pile Less Turned—A Case for 'Low-Input' Composting," CV Compost, updated January 2015, https://www.cvcompost.com/general-composting-guidelines-and -articles.php.

CHAPTER 4: SUCCESSFUL SEED STARTING

1. Taiichi Ohno, *Toyota Production System: Beyond Large-Scale Production* (Portland, OR: Productivity Press, 1988), 123.

CHAPTER 8: WEED AND PEST CONTROL—WITHOUT *MUDA*

1. The practice is described in Jean-Martin Fortier, *The Market Gardener: A Successful Grower's Handbook for Small-Scale Organic Farming* (Gabriola Island, BC: New Society Publishers, 2014).

CHAPTER 10: LEAN APPLIED TO OUR BEST-SELLING CROPS: SEVEN CASE STUDIES

1. Mike Rother, *Toyota Kata: Managing People for Improvement, Adaptiveness, and Superior Results* (New York: McGraw-Hill, 2010), 6.

2. Sara Bir, "From Poison to Passion: The Secret History of the Tomato," *Modern Farmer*, September 2, 2014, http://modernfarmer.com/2014 /09/poison-pleasure-secret-history-tomato.

3. United States Department of Agriculture, "Food Availability and Consumption," accessed March 26, 2017, https://www.ers.usda.gov /data-products/ag-and-food-statistics-charting-the-essentials/food -availability-and-consumption.

CHAPTER 11: FINDING GOOD LAND

1. Purdue University Agriculture News, "Report: Indiana Farm-Related Fatalities Up 10% in 2015," news release, September 29, 2016, accessed January 30, 2017, https://www.purdue.edu/newsroom/releases016/Q3 /report-indiana-farm-related-fatalities-up-10-in-2015.html.

CHAPTER 12: INFRASTRUCTURE AND FARM LAYOUT

1. Taiichi Ohno, *Taiichi Ohno's Workplace Management: Special 100th Birthday Edition* (New York: McGraw-Hill, 2013), 112.
2. Thomas C. Hubka, *Big House, Little House, Back House, Barn: The Connected Farm Buildings of New England* (Hanover, NH: University Press of New England, 2004), 3–30.

CHAPTER 13: LEANING UP GREENHOUSES

1. Credit goes to Vernon Grubinger, the vegetable and berry specialist for the University of Vermont Extension, for teaching us this tip.

CONCLUSION: THE *KAIZEN* FARM

1. Akira Hayami, *Japan's Industrious Revolution: Economic and Social Transformations in the Early Modern Period* (Tokyo: Springer International, 2015), 99.
2. Adapted from table 6.1 in Hayami, *Japan's Industrious Revolution*. Hayami writes, "This is an aggregate of approximately 700 villages. The figures for the 1670s are from 'Records of Villages in the Kambun Period (Kambun muramura oboegaki)' and those for the 1820 figures from 'Records of Villages in Owari Domain (Gunson junkōki)'" (99).
3. Hayami, *Japan's Industrious Revolution*, 100.
4. Malcolm Gladwell, *Outliers: The Story of Success* (Boston: Little, Brown and Company, 2008), 233.
5. Hayami, *Japan's Industrious Revolution*, 73.
6. Gladwell, *Outliers*, 233.
7. Jeffrey K. Liker, *The Toyota Way: 14 Management Principles from the World's Greatest Manufacturer* (New York: McGraw-Hill, 2004), 4.

APPENDIX 1: TOOLS THAT ADD VALUE

1. Henry David Thoreau, *Walden: or, Life in the Woods* (1854; New York: Penguin, 2012), 133. To be fair, Thoreau paid a hired worker for "plowing, harrowing, and furring," at a cost of $7.50. His note next to that line item: "Too much."

Index

Note: page numbers in *italic* refer to figures.

About the Author

Ben Hartman grew up on a corn and soybean farm in Indiana and graduated college with degrees in English and philosophy. Ben and his wife, Rachel Hershberger, own and operate Clay Bottom Farm in Goshen, Indiana, where they make their living growing and selling specialty crops on less than one acre. The farm has twice won *Edible Michiana*'s Reader's Choice award. *The Lean Farm*, Ben's first book, won the Shingo Institute's prestigious Publication Award. In 2017 Ben was named one of Grist's fifty emerging green leaders in the United States. Clay Bottom Farm has developed an online course in lean farming, which can be found at theleanfarmschool.com.

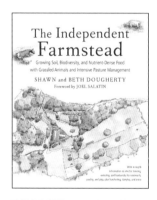
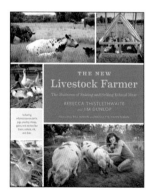

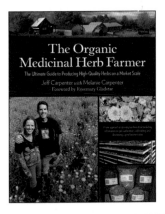
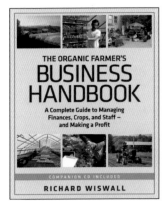
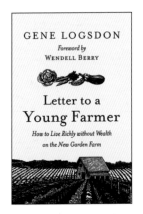
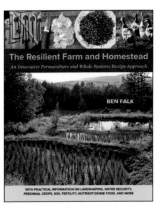